Praise for Scyller Borglum and *STEM Study Habits*

The techniques and information that Scyller offers are an excellent and brilliant roadmap to a student's success. This book is like having Scyller as your own personal mentor cheering you on and rooting for your success. Scyller is an amazing role model, and the details provided reflect Scyller's professionalism and continued commitment to mentoring young people.

—DEBBIE E.
Mom of Two Former Students

My most obvious setback was during my third year in college. I have always been a straight-A student, and suddenly, my grades were barely high enough to keep me in college. I received two warning letters, and was on probation for my low GPA.

It was then that I met Scyller Borglum, who was a master's student in the college of engineering and natural sciences. I told Scyller my problem, and Scyller was kind enough to agree to help me. In my first meeting with Scyller, these words stuck with me: "Forget the past, remove all the noise, and buy a new binder!" I did.

Scyller walked me through all of my study habits, helping me get rid of the bad ones and replace them with new and good ones.

I was suddenly able to look further in the future. I started to plan ahead and effectively manage my time. During that semester, I had six core classes, all of them 500-level. I finished the semester with five As. My GPA was a 3.8 out of 4.0.

—HASAN A.
Former Student

As long as I have known Scyller, Scyller has stood at the doorway of the STEM field inviting, encouraging, and supporting all who consider entering. Scyller has shown that the field is truly open to anyone and the skills learned therein can establish a foundation of success, regardless of one's future in education, business, or even politics.

—MICHAEL R.
Former Student

Unlock your potential in STEM with effective study habits. This comprehensive guide is filled with invaluable tips and strategies to cultivate effective study practices. Through this essential resource, you will develop the strategies and techniques that lead to academic success in the ever-evolving world of STEM.

—MARY N.A.
Retired Engineering Professor

STEM STUDY HABITS

SCYLLER BORGLUM

ONYX PUBLISHING

First published in 2024 by Onyx Publishing, an imprint of Notebook Group Limited, Arden House, Deepdale Business Park, Bakewell, Derbyshire, DE45 1GT.

www.onyxpublishing.com
ISBN: 9781913206697

A CIP catalogue record for this book is available from the British Library.

Typeset by Onyx Publishing of Notebook Group Limited.

To my Students

Contents

Preface

Take all the courses in your curriculum. Do the research. Ask questions.
Find someone doing what you are interested in! Be curious!
—KATHERINE JOHNSON

Dear Student,

This whole journey started with me dropping the keys for my company vehicle, along with my cut up corporate cards and resignation letter, in an overnight envelope. The decision had been made: after nearly five years working as a specialty pharmaceutical sales representative, I was going back to school to study engineering.

But let's go back to the beginning.

Right after graduating from Pacific Lutheran University (where I studied international business with a minor in Scandinavian studies), I moved to Norway for a year on a Fulbright Scholarship. My thesis topic: Sustainable Development through the Venue of International Business. It was there, in Norway, that I fell in love with the energy industry. After countless interviews and tireless research about their oil and gas businesses, I wanted into that exciting world. The problem was, getting into the energy industry can be nearly impossible if you don't already come with a technical background. I moved back to the States assuming I would attend law school and make my way through those tough-to-open corporate doors as an oil and gas attorney. Several fine (but not spectacular) LSAT scores later, I lamented my story to my friend (who by this time was tired of my lamenting).

"Why don't you study petroleum engineering?" she finally suggested.

I had never even heard of this degree.

"Well, weren't you good at math and science in high school?" she shrugged.

Yes, I'd earned As and Bs, but that had been more than fifteen years before.

I pushed the suggestion to the back of my mind.

Still committed to the idea of being an attorney, I visited an admissions counselor, who recommended that I start with at least Calc 1 and geology. Those two courses could be taken at a technical college and transferred into a degree program, should I decide to continue.

I took these courses completely convinced that I would fail both and then make my way back to law school.

I earned an A in one and a B in the other.

Suddenly, I realized, *I could really do this. I could go study engineering!*

The story of my adventure from Texas to Montana I will save for another book. What I will tell you is this: if you show up, work hard, ask questions, and commit to the finish line, you can earn a STEM degree.

Yes, *you* can earn a STEM degree.

This is the book I wish someone had given to me when I went back to school to study engineering. When I set foot on Montana Tech's campus to study petroleum engineering, I quickly realized that transitioning from sales professional to STEM student was going to be as abrupt as it was brutal. What's more, it became apparent that the haphazard study habits I had acquired during my first college go-around were not going to cut it this time.

In high school, I'd earned decent grades, though not through studying: like many of my peers, I eventually mentored and tutored, had excellent memory recall, paid enough attention in class to get the gist of the material, and could reason my way through the rest.

Upon my return to school as a post-baccalaureate, the only classes I took were math, science, and engineering courses, and I planned to get through them as quickly as possible so I could return to the professional world. This meant taking as many as twenty credits a semester, with no "breather" courses. No phys ed. No English or history. No art. Just nonstop technical classes and their corresponding labs.

It didn't take me long to realize that my prior method of "show up and pay attention" (with maybe one late night of studying thrown in there for good measure) wouldn't be enough. In fact, the very first grade I received was lower than any grade I'd ever gotten in my life.

This was a sign: something needed to change. I'd given up a lucrative and established career in specialty pharmaceutical sales to study petroleum engineering, so if I was going to take three to four years out of the workforce to re-tool for an entirely different industry, I needed to do so from the ground up, and I needed to do so properly.

I started by picking out the students in my classes who asked the best questions and received the highest marks and asking them how they studied (and if I could study with them). But while each student had one or two good pieces of study advice, no one had the full package.

The lifestyle habits they lauded I already had a decent handle on (courtesy of my prior bachelor's and master's degrees). The study habits were new to me.[1]

Keep in mind that I went back to school for engineering at the age of thirty-two, which meant that most of my classmates (the ones whose wisdom and experience I was seeking) were at least a decade younger than me. This book is as much a tribute to their generosity as it is a compilation of my own field-tested study habits.

Learning means being willing to set aside our current comfort for future competency. It's not easy to ask for help, never mind admit when we're not doing as well as we think we should be. That's why I wrote this book, my friend. We're not going to get it right straight out of the gate, but your willingness to learn—to fail forward—is exactly the attitude you need to succeed in your degree program.

It is difficult—I know it is—so before you throw in the towel, use it to wipe your face first, and then read this book. You were smart enough to sign up for a STEM degree in the first place and wise enough to seek out and pick up a book that just might give you the tools you need to get to the finish line. Plus, I know there's some success somewhere in your background that

---

[1] For the record, we will be covering all these lifestyle and study habits herein. No stone left unturned!

you can look to for a reminder of the fact that you are capable of doing the hard work that it takes to finish something worthy of finishing.

While I never planned to earn five post-secondary degrees, that is where I landed. After I finished my MS in petroleum engineering, I started my full-time position as a production engineer in the Bakken, North Dakota. The company who hired me knew I wanted to earn my PhD, so we made arrangements for me to commute back and forth between Dickinson, ND, and Rapid City, SD, both to work full-time as a production engineer and to attend class full-time for my doctorate in geology/geological engineering.

Less than a year into my career, the oil boom busted, and I moved to Rapid City to finish my degree in short order. For a few reasons (again for another book), I thought I might need my PE (Professional Engineer's License) in order to move forward in my career. But in South Dakota, you need a bachelor's degree in engineering (not a master's or a doctorate), so once again, I found myself on the road commuting this time between Butte, MT, and Rapid City, SD, completing two different degrees in two different disciplines in two different states. In this way, I finished my (now-second) bachelor's degree, and then my doctorate.

I tell you this zigzag journey so you know that regardless of whether you are still in high school or decades have passed since your college graduation (or when your college graduation would have been, if you had pursued STEM immediately after high school), a STEM degree is an incredible way to begin, continue, or advance your career.

I am a second career engineer who has tutored many incoming freshmen and some sophomores, and the story is almost always the same: "I was fine in high school, but I'm drowning now." If this is you, do not give up. Take this book and mark it up instead. Focus on one new habit a week to get yourself back on track. You do not need to earn straight A's. And guess what? You will have an exciting career no matter what type of STEM discipline you study. And guess what (again)? You are needed! There are more technical careers available than there are people ready to fill them. You are *desperately* needed, and greatly valued.

But enough with the introduction. What I'm getting at is, you've made it this far, so let's get to the finish line together. When you do, flip to the back of this book and make note of my contact information. I want to hear about your incredible success and the wonderful opportunities that are just ahead of you! The world needs your smarts, tenacity, and creativity, and I want to hear your story.

Always,

Dr. Scyller Borglum

1

The Habits You Build in College

We don't all need to be told, but we do all need to be reminded.
—ZIG ZIGLAR

T HE HABITS YOU BUILD IN college are the habits you keep for life. If you were going to buy a house or a new vehicle, you would put some research into it, and you would likely find out what kinds of routines and maintenance you would need to integrate into your life to keep it in top condition. College degrees are incredibly (and similarly) expensive, which means it makes sense to treat your time in college with the same level of care and attention that you would with any other expensive investment.

College should also be a lot of fun. If you are constantly stressed out and tearing your hair out over assignments and exams, then making your way through this labyrinth won't be remotely enjoyable. You can have a very successful college career and go on to a professional career, the military, or (should you choose) graduate school to continue your research, all while enjoying the process.

Yes, really.

So you can make the most out of these four or five years that you have as an undergrad in your chosen STEM field(s), I have assembled all of the lessons I have learned and wisdom I have gleaned from the many mistakes I made as a student over the course of my grand total of thirty-two semesters in school. If I can help you skip over many of the traps, shortcomings, and failures that I encountered during my time in school, your time over these eight to ten semesters should be much more effective. We will have you high stepping across the stage when your tassel is moved from right to left!

You will not need to master everything in this book, but I expect there will be a few things that stand out to you personally. Those things will help you go about college life just a little bit better and a little bit more effectively, so that you can go out into this world and help to solve some of these incredible challenges that we have right at our feet.

As I mentioned in the Introduction, I did not follow any set study method or routine while I was in high school. I was smart enough to pass tests without really studying for them, I did homework as required, and I was incredibly involved in many activities (because I found them more interesting than my studies). As the icing on the I-don't-need-to-study cake, I never saw myself as somebody who would pursue multiple post-secondary degrees. In other words, I did not see myself as someone who needed to develop good study habits.

When you are in a STEM degree program, you cannot afford to approach your studies with that kind of glib attitude. You first need to see yourself as someone who practices good habits. A future engineer does not stay up partying when she has homework due the next day. A future scientist does not turn in sloppy homework or slovenly lab reports because he could not be bothered to write them correctly. A future mathematician does not skip practice problems because math rep(etition)s are too time consuming. No: an engineer, a scientist, a mathematician, first sees themselves as someone who takes great care in their craft, and then puts this care into practice.

A habit can be either good or bad. We are all acutely aware of our bad habits, and we often have the goal of building new good habits, usually around the first day of the year. There are scores of books written on habits, and this book specifically deals with the habits that will serve you well in college and beyond, whether you choose to do postgraduate study or not.

There will not be anything revolutionary discussed, but if you employ most (or all) of these good habits, you will experience tremendous success in college and go on to have the kind of life you are working so hard to have.

So, let's jump in and talk about what it takes to build a habit before we talk about the habits you will need to build.

If you picked this book up because you just received a grade on an exam that you have never seen before in your life, this is the book for you. Every habit in each of the upcoming chapters is transferable to your grand adventure after college. Build, use, and carry these habits with you, and live your best possible life.

Let's get started!

What Are Habits?

Habits are the premade decisions and actions we rely on in everyday life so that we do not have to re-decide every decision needing to be made. If you have your alarm clock set for 5:05AM every morning, then you do not have to decide every night what time you are going to get up in the morning. The decision has been made and your alarm is already set. If you already know that you are going to be on time for every class and meeting that you attend, then you do not have to decide at the last minute if you are going to leave too late and have to call and make excuses, or if you are simply going to leave a shade earlier than needed and always be there on time. If you decide that you are not going to smoke or drink alcohol in college and that you are only going to put healthier options in your body, then you do not have to re-decide every time the offer of wine and cigarettes is presented to you: you can graciously decline and grab a club soda and a lime.

How to Build a Habit

Building good habits is not easy, but nor is living with the consequences of bad habits. So, if you've picked up this book and have read this far, decide now that you are going to a) finish your degree and b) build good habits, because the habits you build in college are the habits you keep for life. We only get one shot at it,[2] so we might as well work hard and enjoy the rewards.

Aristotle said, "It is frequent repetition that produces a natural tendency," and millennia later, Zig Ziglar followed with, "Repetition is the mother of learning [and] the father of action, which makes it the architect of accomplishment." Basically, if we want to make something a natural part of who we are so that we can go on to accomplish great things in life, this starts with repetition—and, of course, this applies to building a *bad* habit, too. Some habits take three weeks to build, and some take three years (not three years of *saying* we'll do the habit, but three years of *repetitions*). For our purposes, we want to build good habits and possibly eliminate a few of the bad ones. That is where our focus will stay over the course of this book.

Everything in this book is going to be about repetitive action. Consistent, day in, day out, boring, predictable, you-can-count-on-it repetition. Most of the successful people I know and admire have the most "samey" routines and daily lives I've ever heard of. They decided somewhere along the line that they wanted to be the best at what they do, and successful people do not squander their time and creative energy remaking the same decisions hundreds of times. Successful people get up every day having already decided how they will behave, how they plan to attack the day, what they need to do in the event of any contingencies, and who they will and will not spend their time with. In this way, they proceed through their day in a healthy manner and with a positive attitude.

I will note here that when I talk about "successful people", I'm sometimes talking about people with a great deal of money and fancy titles, but more often than not, I'm talking about those who have built a life for themselves and their families that supports them in the way they want to be supported and allows them to pursue the interesting things they want to pursue, title or no title. To become one of these people, we must first put into practice

---

[2] Life, that is. College can be done many times over, if desired—as demonstrated by yours truly!

many good habits concerning our daily routines and rituals—habits concerning our physical, mental, and spiritual health; concerning how we interact with others; concerning how we present ourselves to others; concerning our ability to communicate. All these areas are covered in this book, and I can promise you two things: first, you are going to have to practice these habits every single day; and second, you will drop the ball from time to time. Nobody gets it right every time, all the time.

The more you go after each individual goal (be it waking up early, making your bed every day, or turning in homework neat and on time) and practice the accompanying steps day after day after day after day, you will eventually find that it truly becomes a habit. You will follow through on your habit without thinking about it and then move onto the next one.

The more you build these habits, the more you will have a life about which you can smile and be proud. What's more, you will also be building the future life that you always wanted, and that, my friend, is success.

Now, let's look at how we can build the specific habits we want to, so that we can have the life we never knew we always wanted.

Habit by Brute Force

While James Clear describes the "cue, craving, response, reward" method for initiating a habit change, Jocko Willink takes more of a "brute force" approach. To paraphrase Willinck: "How do I get up earlier? I get up earlier. How do I work out more? I go work out more. How do I stop thinking about my ex? I stop thinking about my ex." Not a lot of thinking, researching, dissecting, or reflecting is involved. Just do it, or just don't do it.

I find that both these methods (and sometimes a hybrid of the two) really work for me. In general, when I am modifying a habit I already have (adding ten push-ups onto the end of every workout, for example), I only need to think about a cue to get me started (Clear's method). However, if it is a brand-new habit and one that I have nothing to connect it to or "stack onto", more brute force is required (Jocko's method). Working on this book, for example, fell into the "brute force" category. My workdays are chaotic, so I booked time to write and edit every day. Curveballs arose every so often—curveballs that meant I couldn't always work on it when I'd planned to—so at the bare minimum, I spent the last twenty minutes of my day writing and editing. It was not pretty, but I "brute forced" it and got it done.

No matter what method or blend of methods you use to create and maintain your habits, remember this: discipline is what separates the big dogs from the puppies. There most certainly will be days when you do not want to do your homework, go to class, take a shower, train, memorize another equation, or go to bed on time. The dailiness of good habits is not sexy; it is mundane. But it will see you through to a strong finish, in college and in life.

We Show Up Daily

I like to give my New Year's resolutions a head start. This means I actually start putting whatever fitness, financial, or intellectual goals I've made into practice around Thanksgiving.

On Thanksgiving Day a few years ago—what had always been a day of elated celebration for most of my life—I found myself underwhelmed by the day I'd had. Historically, I could (and would) spend this day giving thanks for my health, employment, family, friends, interesting studies, comfortable home, trusty dog, healthful food, fresh air, wide open spaces, and densely foliaged parks. Most and best of all, I would give thanks for the future: possibility and opportunity. That year, though, the future did not fill me with sunny optimism—but it did not fill me with wintery dread, either. Instead, I found myself looking forward with an arms-folded-across-my-chest, squinty-eyed wariness: "Let's just see how this thing shakes out, shall we?"

This kind of world-wariness does not sit well with me, precisely because it is un-American. I am sure there are some hold-back, do-nothing Americans out there, but I personally do not know any. The Americans I know and love get involved. They engage. They push and retreat and then go sideways, up, over, around, under, and through. They throw, toss, catch, hug, release, jump, crawl, walk, sprint, inch, and lunge. They tweet, call, message, text, sing, type, print, hunt and peck, chicken scratch, and say hello. They show up. This is why, on that Thanksgiving Day years ago, when I recognized my lack of enthusiasm, I gave thanks to Benjamin Franklin.

Benjamin Franklin was born in Boston under one governmental regime, and he died in Philadelphia under another. He was the youngest son (and number eleven of thirteen children), a polymath, and a leading author, printer, political theorist, politician, freemason, postmaster, scientist, inventor, civic activist, statesman, and diplomat. I finished *The Autobiography of Benjamin Franklin* the day before Thanksgiving and grew to love this engaging and engaged, boots-in-the-mud American. He ran away from home while he was young (his dad indentured him in servitude to one of his older brothers), and so he did not have even a surname to recommend him when he entered the "real world", nor did he have any money. He knew he lacked education, so he set about self-cultivation: he read and studied books, taught himself three additional languages, and worked. Almost as an aside, Benjamin comments on his contributions to building the local hospital, an orphanage (in Georgia, of all places), and a library.

In short, Benjamin Franklin showed up. He scheduled his day and practiced his habits. Clearly, I recommend his autobiography. He lists out all the habits that contributed to his success there, and I strongly recommend you peruse this list.

I give thanks to Benjamin Franklin for showing up; for living as a humorous, flawed, deliberate model for Americans three centuries hence. Because then, on that Thanksgiving day when I faced the (drab-looking) future, I was given a clear road to success and fulfilment: show up, consistently and especially when you don't feel like it or the odds are not in your favor. Because actually, they are.

Recommended Reading

- *The Power of Habit: Why We Do What We Do in Life and Business* by Charles Duhigg.
- Atomic *Habits: Tiny Changes, Remarkable Results* by James Clear.
- *Discipline Equals Freedom: Field Manual* by Jocko Willink.
- *Discipline Is Destiny: The Power of Self Control* by Ryan Holiday.
- *The Autobiography of Benjamin Franklin* by Benjamin Franklin.

Strategies for Success: Habits

- At some point, your motivation will die off, so rely on your hard-won habits to keep you going.
- Do not try to change all your habits at once. Pick one or two that will make the biggest impact on your current situation, and develop those until they become routine. Then, move onto the next habit.
- Treat your time in college like the training that it is. You are going to be a professional whose skills, creativity, and abilities will help solve the problems of our age, so train like the professional that you are.

2

High School versus College

The most difficult thing is the decision to act. The rest is merely tenacity.
—AMELIA EARHART

THERE ARE THREE BIG DIFFERENCES between high school and college.

First: in college, you are responsible for driving yourself to the finish line and to a life where you are responsible for... everything. When you were in high school, you had the same class five days a week and straight through the day. In college, however, you will have a class either on Monday, Wednesday, and Friday for fifty minutes, or on Tuesday and Thursday for about an hour and a half.

You will notice that there are large chunks of commitment-free time in your college schedule, but do not be lulled into a false sense of security: expectations regarding what you do in your free time increase significantly in college. While in high school you might have had time in class to do homework and review materials, this will not be the case in college. All extra assignments (with the exception of some of your lab work), including your homework, problem solving, and report writing, plus any necessary studying, will be done outside class in your own time. For the most part, you will have lectures when you are in class (no "homework completion" time), the homework assignments will take longer, and you will have more in depth concepts to work on outside of class. How you choose to organize your time is entirely up to you, so choose wisely (see Chapter 6).

Second: you have more freedom in choosing your courses. Engineering might be the exception for this, though, as engineering courses generally have a pretty tight curriculum flow path that you will need to follow in order to finish in a timely manner. Even still, you will have more opportunities for electives and a minor, if you so choose.

I recommend not making too many commitments during your first week in the semester: the first week will make you think you will have more free time over the course of the school year than you actually will. Week one is generally known as "syllabus week", and is (you guessed it) when professors tend to hand out their syllabi. It is a shortened week, and the classes are not designed to provide a lot of lecture material or homework. You will therefore think you have all the time in the world based on this one week. Two weeks into the semester, however, reality will hit, and the homework needing to be done and the exams coming up will suddenly necessitate a great deal of work.

Third: in high school, if you don't show up, somebody calls your parents. In college, nobody calls. Attend class, don't attend class... the decision is entirely yours. The catch is, you are paying for each of the lectures in the syllabus, so it is undoubtedly beneficial for you to attend class. If the lectures did not matter, professors would not give them. If you are in a pinch, it is fine to borrow a colleague's notes from a missed lecture, but you never know what they might have missed or spaced out on, and in the end, your professor will have a good idea of whether you have been attending class. If you chronically skip or sleep through class, you'll have a very low chance of passing, never mind learning, the material.

In a nutshell, the responsibility is on your shoulders now. You are an adult, and are expected to manage your schedule like one. College is the training ground for your professional life, and just as you will not have your work schedule dictated hour by hour in the professional world, you will not have your collegiate schedule dictated hour by hour during your remaining time as a student. With tremendous freedom comes equal responsibility.

To prepare you for this transition, we will, in upcoming chapters, explore in lots of detail the important rules I have introduced here that will see you through college life.

Decide Now to Finish

Taking on a STEM degree is similar to competing in an endurance event. It is not a sprint. You will have highs and lows, and there will be plenty of opportunities along the way to celebrate little victories.

When taking on an endurance competition, you pre-decide to finish the race. You need to have this certainty ahead of time so you can make it through the physical punishment involved. You need this same certainty with your STEM degree: you need to pre-decide to finish the degree. If you dabble in it just to "see how it goes" or to evaluate whether you like it, then you will not finish. At some point, the going will get really tough, and you will be beyond frustrated, wondering why you signed up for this level of cerebral punishment.

You will see throughout the course of this book that many of the habits competitive athletes adopt will also be important for you in your STEM training. You will need to take care of your health (physical, mental, and spiritual). You will also need to cement the habits that will train you for professional life and work on your technical skills so that you do well in your classes.

So, before you just wander into a class so you can casually evaluate whether it seems to be a good fit, trust what I am saying here: you need to decide *now* that you're committed to finishing the thing. There will be parties, and you will skip, but don't worry about this, because you *will* get up early to finish a homework assignment, and you *will* work on your organizational and time management skills, and after four to five years of rigorous training, you *will* come out of your degree program ready to help solve some of the world's greatest challenges. But you will *not* be bestowed this honor if you meander through your degree program. Attack with confidence and pre-decide to finish.

Make New Friends

We will spend a lot more time on this subject in Chapter 10, but I wanted to give you a heads up about this at the beginning of the book, because making new friends really is of utmost importance. In high school, you may have had friends you'd had for years—some even going back to elementary school—so now is your chance to branch out and develop a new circle that shares your enthusiasm for your classes.

I love the start of a new school year: fresh faces, imminent adventures, a new T-shirt from the bookstore, making new to-do calendars, buying new used books, warming up my mechanical pencil and resurrecting hallway friendships...[3] "You don't build a relationship the day you need it," as President Lyndon B. Johnson said, so start making those friends today.

Stretch Yourself, But Don't Snap

A crackling, dancing, high-voltage livewire has nothing on my excitement for a new school year. With so many opportunities and adventures to be

---

[3] Hallway friends are those individuals whom you recognize and could not address by name if your life depended on it. Their importance comes in during the cold, dark winter months when you arrive on campus in the dark, leave in the dark, have no natural light in your office, and begin to wonder if you are still part of the material world or if perhaps your experiment caused you to vaporize and leave this familiar plane of existence. If said hallway friend still nods in your direction when they pass you, life persists. Say hello and introduce yourself. Your hallway friend just might become your lifelong friend. If not, no worries: you don't need to be everyone's best friend. The more you say hi to people whom you pass in the hallway, the better you will become at expanding your friendship circle.

had, an infinitely excited student needs to be kept sane *and* still engaged in their academic community, to prevent them from running headfirst into burnout. But how?

These are hard-fought lessons, my friends.

First, understand that saying yes to one thing means saying no to something else. I do not have an addictive personality, but I do have an excessive one. My advisor at MT Tech once said offhandedly that "if Scyller's doing something, you know she's overdoing it".

Point taken.

Second (and this was a hard one for me): understand that there is a difference between a productive student and an effective student (see Chapter 4).

In line with these points (and the power of hindsight), I strongly recommend the following:

- Join student organizations, but not too many. At one point, I belonged to five separate student organizations and held officer positions in two. Too much. However, I certainly recommend joining the clubs that are of interest to you. We'll explore professional societies later in this book, as these introduce students to professionals in their fields, career opportunities, and scholarships.

- Be wary of course overload. Part of my heavy credit load as a post-bacc stemmed from me trying to get through general courses so I could get into the graduate program. I maxed out my semester courseload, and predictably, my grades became harder to maintain. I am still partial to adopting a heavy courseload (because I enjoy learning), but remember that the tradeoff for this will always be grades.

- Don't overwork to earn money (if you can help it). Full-time or part-time school is expensive, and there is no shortage of need for cash, so just work enough to cover the bills and that's it. Get through your program and then get onto making real money.

- Be intentional with your socializing. You have three options, from which you can pick two: being well-rested, having top marks, and having a rocking social life. Choose wisely.

Responsibility

The other substantial difference between high school and college (we alluded to this before) is your expected level of responsibility. These years transition you from a spectator of life, whose feeding, care, and daily activities are administered by a parental figure, to an active participant in life, whose feeding, care, and daily activities are administered by you.

Now, the good news is that this is not *Lord of the Flies*: you have not been thrown out into the world completely alone, and life in college is far from nasty, brutish, and short. Plus, you do not yet have to be responsible for *everything*. There are groups on campus who are there to help you. They are not there to push you through, but they are available to make sure you learn your material, and can help you navigate your way through this sometimes-complex system. For example, most campuses have tutoring areas that are free to use. They hire students in a variety of subjects, and you schedule a time that works for you and your tutor. Here, you can learn anything and everything, from chemistry to calculus, physics to physical health sciences, and business finance to economics.

You, however, will be responsible for the "bigger picture" items: managing your time; getting to class on time (and attending!); turning in your homework assignments without being reminded; preparing for exams and end-of-semester projects; keeping yourself healthy; maintaining something of a social life; taking courses in the order in which they need to be taken so you can graduate in four or five years; and so on. Depending on your circumstances, you will also be responsible for managing your money, keeping your dorm room (or rented house or apartment) clean, paying your bills, and attending to any leadership positions you might have taken on in organizations.

These are all good exercises for strengthening your "independence" muscles.

In high school, you likely lived with your parents (or certainly a guardian), all the bills were paid, your food and clothing were given to you, and your classes were pretty well laid out. With the increased freedom of college, you certainly have increased responsibility. This is where you can really shine!

Course Behavior

Courses in college are, unlike high school, essentially big meetings: you have one person standing at the front of the room, but instead of a manager or a supervisor, it's your professor or TA. When you are in the classroom, maintain the same level of respect and good manners that you would employ in a professional meeting. This means sitting up and at attention, with your notebook or tablet in front of you. Do not sit there with one earphone connected to your phone listening to a game or a podcast while keeping an ear out to see if your professor says anything you deem to be of consequence. You are not so clever that you will be able to pass a class in this way. And yes, your class leader will notice.

Make sure your phone is put away, and do not use it during class. Classes are generally fifty minutes long on a Monday, Wednesday, or Friday, and an hour and a half on a Tuesday or Thursday, and occasionally, there will be specialty seminars that run a bit longer. Either way, you can go that long without checking your phone.

Do not bring noisy food and drink into the classroom. If you bring in a water bottle or a cup of coffee, that's one thing, but if you are slurping down a soda with ice in the bottom of a cup or crinkling your hamburger wrapper, that is exasperating. Be courteous to your classmates. You do not want to be a distraction to those around you. They paid for this class just like you did.

If you start coughing, choking, or sneezing uncontrollably, discreetly step outside and finish your business. Do not chortle or hack your mucus in the back of your throat. No one is impressed with this demonstration of laryngitical exhibitionism.

When you are in class, if your professor takes questions and you have one during the lecture, address your class leader respectfully. If, conversely, the culture is such that you do not ask questions during class, wait until the end, and then address your class leader directly. Remember, you are in training for your career after you graduate, so it is important to be able to read the room.

Do not use foul or coarse language in the classroom. Do not be disrespectful. Attending class and having your work done ahead of time will demonstrate to your class leader that you take it seriously. The day may come when you need a letter of recommendation from them, so don't burn bridges unnecessarily, and certainly not before they are built.

Whether you are planning on joining the military, the professional world, or graduate school after your degree, you are in training for the real world every day that you are a college undergraduate for, so take the opportunity to practice and reinforce good habits, including good manners.

How Not to Interact with Your Teaching Assistant

One semester, when I TA'ed a lab, I had a student who routinely showed up late. He would wander in as I was lecturing and then start talking to me mid-lecture, as though we were old friends catching up on gossip or the day's events. I was completely appalled by his behavior, and after a couple of episodes of this, I asked him to come see me after class. I explained to him that he needed to show up on time for class, not interrupt when I was lecturing, get to his seat, and get his work done.

He seemed truly bewildered that this was how one was supposed to behave in a lab setting. To this day, I have no idea if he was truly that clueless about proper course behavior, or if he just wanted to see how long he could get away with it. Either way, it did not cast him in a good light for me.

On another occasion, I had returned graded papers to students in class, and the grades were lower than the students had wanted. After class, a male student who was considerably taller and larger than I was (and I'm certainly not small; I was an open weight rower in college!) was enraged at his grade, and demanded that I account for his score. He kept closing in on me, and as I backed away, I ended up pinned against the wall. He let me know, in no uncertain terms, what he thought of his grade, and he demanded that I reexamine it. I said what I needed to to negotiate my way out of there, made my escape, and reported him to the professor, who I asked to review the assignment in question and then grade all his work going forward. That way, I sidestepped any future interaction with him and prevented any bias from affecting his grade.

Rather than owning his poor study habits or inability to do the work, he took it out on his TA, and, as I found out later, other class leaders. I do not believe he ever graduated.

As a grader, I do not have any issue reexamining someone's work. I make mistakes, and am happy to give points that were unfairly or unduly docked. As a student, I have been on the other side of this, where I believed my work was unfairly and harshly graded, and this is frustrating, especially when you put a great deal of effort into your assignments. Wait a day to have the conversation if you need to, and ensure you do not let your frustration affect what should be a professional conversation. If you continue to have an issue with your grades and believe you are being unfairly scored, first talk to the TA (if they graded your assignment) to determine if you are missing a bigger picture item. If you still believe you are hitting all the assignment requirements and do not believe you have been fairly heard, then talk to the professor.

The last example I have of what *not* to do is about a student who wanted more time to complete an assignment. Rather than asking me for an extension discreetly either in the hallway or during office hours, he bellowed this request across the classroom in front of all the other students. In general, he was a good student with high marks, and if he had asked politely, I probably would have cut him some slack and allowed him a little extra time. Once he made his demand in front of the entire class, though, I had no option but to adhere to my boundaries and say, "No. Time's up."

If you have earned some grace with your TA because you usually turn in your homework assignments neatly, completely, and on time, you are consistently punctual for class, and you generally do a good job, then do not sabotage yourself in the end by making special demands in full view or hearing of everyone else in the class. No TA likes to be bellowed at, and they certainly will not capitulate to such behavior.

Stupid Questions

If you have heard somewhere along the line that there is no such thing as a stupid question, let me correct that for you: yes, there is such thing as a stupid question. Stupid questions fall into three general categories:

1. Not paying attention and asking about something that has already been addressed extensively in the presentation.
2. Posturing. This is when your question is designed to demonstrate how much you know, with little to no regard for the speaker's expertise.
3. Hyper-specific questions that are only relevant to your own agenda. Feel free to ask the speaker later, one on one, if you have specific questions, but do not weary the rest of the audience while you mine the speaker's knowledge depths on a micro-subject for your own satisfaction.

Questions that are *not* stupid are those that help you understand what you are trying to learn, based on what you have already understood from what is being taught, and with appropriate intentions. Period.

Recommended Reading

- *The Coddling of the American Mind: How Good Intentions and Bad Ideas Are Setting Up a Generation for Failure* by Jonathan Haidt and Greg Lukianoff.
- *Fortitude: American Resilience in the Era of Outrage* by Dan Crenshaw.
- *Grit: The Power of Passion and Perseverance* by Angela Duckworth.
- *Lord of the Flies* by William Golding.

Strategies for Success: College

- Decide now to finish your degree.
- Know that you are responsible for your homework, class attendance, and performance.
- Ask for help when you need it.
- Behave in your courses like you would in a professional meeting.
- Treat your TA or professor with respect.
- Don't ask stupid questions.

3

Taking (and Passing) Exams

Everyone has a plan until they get punched in the mouth.
—MIKE TYSON

THERE ARE TIMES WHEN BEING bright is a disadvantage. Most of the bright high school students I meet and work with do not have functional study habits by the time they hit college, because they never really had to study. Strong memory recall and good reasoning skills formed the bulk of their exam-taking prowess. Accordingly, the number one question struggling freshmen (a.k.a., formerly successful high school students) ask me is, "How do you study for exams?" This is a good question, especially if you have never really done it before.

It is important to know that there is a difference between studying for exams and learning the material (though they are not mutually exclusive). Learning the material is something a student does day in, day out, week in, week out. If I am desperately studying the material the night before the exam, I am not learning; I am cramming.

You may find as you go further along in your undergraduate career that exams become less and less frequent and projects and reports become more and more frequent. This shift matters, because your time in school is there to prepare you for the professional world. By their nature, exams are not how you will generally be assessed in your career (professional and licensing exams excepted, of course). Your boss or supervisor is not going to sit you down in a room by yourself with a pencil and calculator and give you a test that you must answer all by yourself. In fact, you will generally be expected to work in teams. Thus, I completely sympathize with how frustrating and artificial exams can feel. For the moment, however, this is the only method professors have of reliably assessing your individual knowledge base and how well the class in general is picking up the material. Departments also use exams as a method for accrediting agencies to judge the academic program. So, exams will likely stick around for the time being.

Before we begin: the most important piece of advice I can give you about taking an exam is *follow the instructions*. Professors are not out to get you— they want you to do well—but on occasion, they will also want to test your ability to read and follow directions. There will be no recourse for you if you do not read and follow instructions.

Now those things are out of the way, let's talk exams.

Types of Exams

There are two types of exams: incremental and comprehensive.

Incremental Exams

Since semesters are usually sixteen weeks long, you can expect an incremental exam once a month, or about every four weeks. This is particularly true in your first two years of undergraduate coursework. Most of the 100- and 200-level courses out there are survey courses that cover a great breadth of material, but not in much depth. Think a mile wide and an inch deep. The purpose of these survey courses is to introduce you to new material that you will study in more depth in more specialized 300- and

400-level courses, when you are an upperclassman.

Because there is so much material and to ensure progress is being made, your professor or class leader will likely test you throughout the semester with periodic incremental exams. Since these exams are typically standalone, you will only need to study for the content that was covered immediately before the exam. Do *not* focus on anything else.

This, contrary to popular belief, does not necessarily make studying easier or more straightforward. If a limited amount of information is covered, this gives your course instructor free rein to ask you very detailed questions about this content.

These exams' format can range from multiple choice, true/false, and fill-in-the-blank, to short essay answers. Most professors know that if they assign a great deal of writing, they will have to grade a great deal of writing, so in larger classes, there usually isn't as much writing to do. Depending on the course, you may also need to identify components of images or maps.

Comprehensive Exams

Comprehensive exams are almost exclusively taken at the end of the semester in the form of a final exam. I say "almost" because there may be a professor out there who routinely gives comprehensive exams throughout the course of the semester, but I personally have never come across one. Comprehensive exams are usually longer, and cover all the content taught in the entirety of your semester.

Professors vary quite a bit in their approach to comprehensive exams. You may have one who allows you to bring absolutely nothing to the exam except a pencil, or you may have one who uses a "take home and use all the materials at your disposal" approach. My best advice to you here is to talk to students who have taken the course a semester or two before you and have had the same professor. Find out what to expect ahead of time so that you can plan as you go.

The harder comprehensive exams are those where you are not allowed to take in any additional material except for your brains and pencil. Knowing whether this will be the case ahead of time will mean you can study along the course of the semester and put a little extra attention into this course at the end of it (the end is going to be a big finish). This is where notecards and daily walks will really help you shine (we'll discuss those later).

If you haven't taken the time to *learn* the material, this style of exam will almost exclusively be about memory work. In survey courses, comprehensive exams will not generally be about thoughtful contemplation and reason; they will be about facts and data, representative problems and formulas, and general information that you learned (or memorized). You will need to demonstrate that you have absorbed the material and can regurgitate it on paper in a comprehensible format. If you have great memory recall, this will help you. Do not take it for granted, though, because the volume of material tested will be more than anything you experienced in high school. Respect the mountain, but don't tremble before it: be confident that you can climb it.

If you are allowed to use notes and handouts, this may initially sound like a godsend, but remember that if you get into the final exam and you have your binder in a disorganized heap and you're facing a time limit, you will burn up your time looking for information rather than answering the questions. If you have studied along the course of the semester, however (even in bits and pieces), you will find that you know where your information is, and you will do well when answering the questions.

Study Materials

Once I am confident that I understand the material, these are the three sources I use to study for exams: homework, in-class notes and work, and end-of-chapter concept problems. A fourth possible option is to look at prior exams—not *study* from prior exams, but *look* at them—to see what is expected. Tread carefully with this one, as some professors consider their prior exams proprietary and grounds for cheating.

How should you attack studying for exams in order of the biggest bang for your buck?

1. Homework. If you are in a time crunch, this is the most efficient place to review for your exam. Professors like to use homework problems for exam problems because a) they are already written (and it is easy to swap different variables in and out) and b) it means students can't come back and say, "We never covered this in class!"

2. In-class notes and work. If the professor saw fit to cover it in class, you can bet that it is fair play for an exam. Grab a fellow student and take turns teaching your way through the slides (use a whiteboard if that helps!). Working on in-class problems will reinforce the material that your professor zipped through but still considers vital.

3. Concept problems. Math, science, and engineering professors all have required texts and give reading assignments, and nearly every textbook I've used has had concept problems at the end of the chapter or book. At the beginning of my doctoral studies, I took a 600-level course on global tectonics, and—you guessed it—there were problems for me to work on at the end of the book. This resource has proven invaluable at every level of study.

4. Past exams. Again, make sure you read the syllabus and know if the professor allows you to have past exams in your possession. If so, then there are good reasons to review past exams. Do *not* study from exams, however. Study from the source material. The benefit of seeing past exams lies in your understanding how the professor asks questions, the number of problems you will need to get through to finish the test, and the level of difficulty compared to homework problems.

Another piece of advice that I love: make sure you understand the material at the end of every lecture. If you don't, sit down with a fellow student, the TA, or your professor, and get it locked in *before* the next lecture. This approach will help you in two ways: 1) you won't be lost before the next lecture begins (incredibly demoralizing!), and 2) you won't have to study nearly as hard prior to your exams (incremental or comprehensive).

As for how much study time is needed prior to an exam, this depends on your comfort level with the material. The more you focus in class, the less time will be needed out of class to study for exams. Go forth and study hard, and then go play.

Exam Nerves

While I have never suffered from test anxiety, I did notice an appreciable uptick in my nerves on exam day during my STEM studies. It could be that my comfort with the material never felt strong, or that the exam halls were enormous and I could hear panic in students' voices... or just that STEM exams are hard. Whatever the reason, the nerves were undeniably there, so I developed a system that kept most of the butterflies flying in formation.

First, I double checked that I had everything I needed for the exam (i.e., a mechanical pencil, extra lead, an eraser, a non-programmable calculator, a formula sheet, and earplugs). Knowing I had all my gear meant no feathers flying; just a cool, calm head to take the exam.

Next, when waiting in the lobby of the building prior to sitting down for the exam, I learned I had to stay away from others and review the material in my own head or on my notecards before stepping in. No matter how well intentioned my peers may have been, the frantic faces around me only worsened my butterflies. As a part of keeping my distance, I wore earplugs for every exam. I put them in as I walked into the building and then would nod if someone smiled or said "hi". I didn't make conversation; just gestured at my notecards and kept going. There is no good that comes from socializing before exams, trust me.

Lastly, I kept breathing. I did not know about box breathing at the time, but I wish I had. Box breathing allows you to regain control if you start to feel yourself sliding toward the edge. To do box breathing, visualize yourself going around the edges of a box: start with three counts on an inhale (up one side), three counts on a hold (across the top), three counts on an exhale (down the other side of the box), and three counts on a hold (across the bottom). Repeat as needed.

Exam Strategies

Your exam starts before you sit down.

You should eat at least one meal before you take an exam. Our brains run on glucose, and if you have not eaten since the night before, you will risk fading mid-exam.

Arrive at the exam room early. Use the time to grab a drink of water, run by the restroom, box breathe (see the previous section), and get your thoughts in order. Locate the handwringers and stay away from them.

When you're doing the exam, start with the problems you know how to do. Skip the ones you are unsure of, or have no idea about. Do not get hung up on a problem early on; save those for the end of the exam. You may not know how to do every problem, but you need to leave nothing blank, so put something—anything—down. Say your prayers to the gods of partial credit, and move on. For multiple choice or true/false questions, do not change your first answer without a really good reason to. Write legibly so the grader can give you the most credit possible.

Have confidence. If you attended class, took notes, and made an honest effort at doing the homework, then know that this is an opportunity to show that you know your material.

Oh, and one last thing: *write your name* (or some personal identifier) *on the exam.*

Recommended Reading

- *Endurance: Shackleton's Incredible Voyage to the Antarctic* by Alfred Lansing.

- *Everything Is Figureoutable* by Marie Forleo.
- *The Little Engine That Could* by Watty Piper.

Strategies for Success: Exams

- If you have less than an hour until the exam, use this time to study your class notes, especially those written by your own hand.
- If you have a few days until the exam, take sixty to ninety minutes a day to review your class notes and re-practice homework problems.
- If you have a week or more until the exam, review your class notes, homework problems, quizzes, and any additional handouts from your professor or TA. Use this strategy for your final exam studies.
- Do not study from past exams.
- Stay away from handwringers on test day.
- Best practice: make sure you understand the material at the end of each lecture. If you do not understand, get after it right away. You will spend more energy putting it off than you will wrestling the material out.

4

Workload: Homework, Lab Reports, and Semester Projects

There are no shortcuts. Everything is reps, reps, reps.
—ARNOLD SCHWARZENEGGER

W HEN YOU ARE COMPLETING YOUR homework assignments, lab reports, and semester projects, what you are really doing is training for the memos, reports, and concept designs you will create for your future supervisor at work. I make this point because it's important that you start practicing the skill of putting work together that is both presentable and respectable now. And yes, you still need to do this while you're juggling time constraints with competing priorities. Again, this is practice for the real world!

Your homework assignments have your name on them, so their quality should be your signature. Your work reflects on you personally. I'm not saying that if you have answers marked wrong that it's a problem; that's a discussion for another section. What I am getting at is (as you will see in a moment), I have had homework assignments turned in to me that bewilder and baffle all sensibilities. So, do not turn in a sloppy assignment. Do not worry about making genuine mistakes, but do present yourself as someone who cares about their product, not as someone who doesn't, or (even worse) as someone who thinks that their genius will somehow shine through the muck and detritus. Take ownership and pride in your work and turn in a worthy assignment.

What does that look like? Let's take a look!

Setup

Everything I am about to share with you stems from my experience grading forty or more labs in one week, many times over. Time and time again, the students who handed in their lab assignments and did well, did very well. Those who handed in assignments and did poorly, did very poorly. For the most part, these students were in their junior year of college, which presupposes the fact that they all passed courses requiring homework—possibly even kindergarten—hence why I wanted to tear my hair out whenever I came across the following:

- There being no name, title, date, and/or pagination on the assignment.
- The problems being completed out of order.
- The name and lab partner names being in the same order on two separate labs for the same lab assignment (meaning I had no idea which lab belonged to which partner).
- Two different versions of the same lab assignment being submitted by one student, with no indication of which version he wanted grading.
- The homework itself looking like it had been run over by a truck (several times).
- Maps (which usually have four boundaries) being drawn with no legend[4] or descriptions, and not to scale. In one baffling map drawing lab, the student turned in a five-boundary map with no explanation or justification. Just a random pentagon! This left me to wonder, *What does*

---

[4] A map legend defines the symbology and/or colors used in the map.

this mean and where on earth is this?!

- Numerical values with no units and no explanation.
- Reports not only being turned in unstapled, but the individual pages being turned in at separate times and to separate lab instructors—as if to say, "I wonder if my TA can figure this little puzzle out!"
- Hand-drawn diagrams spanning five sheets of 8.5" x 11" paper, the sheets taped together, trimmed to fit the diagram, and folded several times.

I graded assignments for years as a TA at two separate universities, and the aforementioned list is based on just one sample (*one*) of the work I reviewed. These grievances are not unique to one class, department, or university; I have seen them be made many times over.

This got me thinking that guidelines for a winning homework assignment might be in order, which would go a little something like this: Write your name, the date, and the assignment title at the top of the first page.

- If the assignment is more than one page long, use page numbers, and preferably like "1/10, 2/10, 3/10", etc. so I know to expect ten pages in the assignment (should one get lost).
- If you worked with a lab partner, clearly note who the author of the assignment is and who the lab partner is.
- Type assignments whenever possible, field notebooks and lab notebooks excepted.

Plus, for engineering or lab assignments:

- Title all figures.
- Give all graphs axis titles, units, and a legend.
- Give maps a title, north arrow (or compass), latitude and longitude, legend, scale, date, and author.
- Caption tables above.
- Caption figures below.
- If you erased and smudged your answers, get a new sheet of paper and rewrite.

- Fit diagrams onto one sheet of 8.5" x 11" paper (unless instructed otherwise).
- Do not turn in an assignment that looks like it has been (or actually has been) run over by a truck (or car, tractor, combine, or golf cart), nor one with a boot print on it.
- Answer the problems in the order they were given in the assignment.
- Staple multipage assignment pages together in order.
- If you tore your answers out of a notebook, clean the fringes off prior to handing in.
- Box answers for engineering and math problems.
- If you are solving science, engineering, or word problems, include units in your answers.

As we have established, your four years in college should be considered training ground for the professional world. If you turn in assignments following these guidelines, not only will your TA feel happy and refreshed when grading your brilliant work, but you will be practicing for success as a professional. No boss wants to review slop, this I promise.

Science

At the beginning of your program (especially in the freshman and sophomore years), your science courses will require a great deal of memory work. The introductory classes are survey courses, and it is in these that you will cover a tremendous amount of material, not deeply but extensively, to introduce you to a variety of subtopics in your discipline.

In an ideal world, you would understand and be able to reason your way through all the concepts, but in my experience, the volume can prove too much and the concepts too foreign for me to fully get on the first pass. Straight up memorization was the only way I could make my way through some of the exams. Accordingly, my recommendation for memorizing is threefold:

1. First, use the extra-large notecards mentioned in Chapter 6's "Tools of the Academic Trade" section. Handwrite the notes you want to memorize on these cards. I generally recommend blue ink, for no other reason than the fact that it is easy to read on a white background and stays fast even when you are in inclement weather.

2. What does inclement weather have to do with it? Well, it turns out we remember better when we combine physical activity with memory work. So, by writing your notes on extra-large cards that you can hang onto and going for a walk (whether it is damp out or not), you can talk to yourself and recite as much as necessary to drill the needed memory work into your brain.

3. When you finish a study session, you may want a quick nap, and this is recommended. There are studies suggesting that naps move information from your short-term to your long-term memory, and that is exactly where you want it.

When you move into upper-level classes for your science degree, you will find the focus to be mainly on using previously learned information to solve real-world problems. This is where the courses really become fun: these problems depend on your creativity and experience. You will not be able to complete these, however, if you do not have the fundamentals ingrained in your thinking. To ingrain this knowledge, you will need put the time in with your freshman- and sophomore-level classes. So, do not skimp on your study time during these years. The more effort you put in then, the more you will be able to spend your time in your junior and senior years developing your problem-solving skills—and *that* is when your studies really get fun.

Engineering

The good news about engineering courses is that they're all rooted in physics. This means that if you took your required Physics 1 and 2 courses (and, in some cases, like in my undergraduate coursework, Physics 3), you will be well-versed in all the principles your engineering courses will teach and test you on. If you did not quite understand the concepts in your physics courses, it is worth putting extra time and effort into studying them, because you will see them again at some point in your engineering studies. All of those physics principles will come around again in the form of engineering problems.

Depending on your engineering specialty, you can count on taking a standard suite of courses—namely, statics, dynamics, fluids, mechanics of materials, and thermodynamics. Thermodynamics is deeply rooted in physics and chemistry, and you will likely take it toward the end of your degree program. Don't be intimidated by the name or title: it is actually a fun class, and you may like it so much that you go on to take advanced thermodynamics.

There are numerous specialty engineering courses that will be tailored to your specific degree program. Even if you go into one of the more specialized engineering disciplines, it all comes back to physics. So, study your heart out in physics, and you will never have to study that hard again.

There are a limited number of problems for and ways in which you can be tested on your engineering acumen. The good news is that if you go through your engineering homework problems, work them out, and then rework a sample of them in preparation for your exam, you will be in pretty good shape to succeed, in college and beyond. If you are struggling, that doesn't mean you should give up and change majors. Many of these concepts are difficult the first, second, and even third time through. Enlist a peer who understands the material, or camp out at your on-campus tutoring center. Either way, keep working the problems until they make sense. It is all about reps, my friend.

While there will not be much in the way of memory work for engineering problems, you will have to know the formulas for solving problems. Generally, the necessary formulas are provided on exams themselves, but not always, so check ahead of time. Your professors will want you to demonstrate that you know how to use these equations to solve the problems, so use what they go over in class when completing your homework and just keep practicing.

As odd as it sounds, I also recommend that you do the assigned reading in your course textbooks. Hearing the same material presented in two or three different ways will help speed up your comprehension and drill the content into your memory.

Math

Studying math is not entirely different from studying a foreign language. Think of multiplication and division tables as vocabulary, algebra and trigonometry as sentence structure, and calculus as writing paragraphs. Each area of math builds upon the skills that have been previously developed (like in a language course), and the great news is that the time you invest on the foundational levels only serves to increase your "fluency" later. Also, just like when you're imitating a native speaker when learning a new language, you need to have examples to follow (from a textbook, notes from class, or online resources) when learning math.

Is there a wrong way to study math? When I returned to college after a fifteen-year hiatus from academic math problems, I found it incredibly frustrating that the guidance we received in this regard was, "Just study it." Fine, but *how*, exactly? In class, I followed along with the professor as she explained every step while writing on the board. Then, prior to the exam, I glanced at my notes to "study". Then, on test day, I choked.

Of course I did.

I could not solve the problems on the exam because... well, I had never practiced them on my own. I equated "following along" with comprehension, when actually, listening to the math professor review problems and reading through my notes was not the same thing as actually solving calculus problems.

So, what's a better way to study math? After studying with top students in my class and asking my professors for guidance, I have learned the following steps, and they have improved my math fluency. It is the reps method. It goes like this:

1. Write down sample problems from class in your notes.
2. Work through the homework problems you have been assigned.
3. Pick out three or four representative problem types and rework, using notes for guidance.
4. Set aside for a period of time.
5. Either later that evening or the next morning, rework the same three or four problems and see if you can do it without assistance from your notes.
6. Made it through? Good to go! Needed to check your notes again? Not a problem, but keep at it until you can work through the problems without assistance.

Initially, it may take up to three hours a day to study math, but over time, you will move through problems faster as the material starts building on itself. Do not be afraid/embarrassed/intimidated to ask for help, and if your instructor is not able to explain it in a way you understand, ask a fellow student, a tutor, or another math instructor.

Finally, keep practicing. Arnold Schwarzenegger has the best attitude toward reps out of anyone I have ever read about. It did not matter if he was lifting or preparing to give a speech to the United Nations, he put in his time with reps. And clearly, it paid off! So, go for it and get your math reps done.

Writing

There are two primary ways in which you can communicate your brilliant discoveries about and understandings of STEM principles to the general public (or your colleagues, or your supervisor). One is writing, and the other is speaking. We will address public speaking in a later section, so for now, let's talk a bit about writing.

Technical writing is different to any other kind of writing that you will ever create or read. This style of writing takes time to learn, and because it is so different from the kind of writing style you probably developed in high school, a fair amount of practice is needed for you to really gain the skills for it.

As challenging as technical writing can be, I happen to like the process. It forces me to work through an entire line of thinking. If there is any gap in my understanding, writing exposes it. Sadly, however, I frequently hear from my fellow engineers how difficult writing is for our kind, and I think this is the case for a couple of reasons.

First, technical writing is challenging because it is uncommon. People write similarly to the writing they read—our reading habits inform our writing style—and in my case, it was not until I had been reading and writing for thirty years that I started actually reading technical papers, never mind writing them. No wonder it was tricky!

Second, like all skills, writing takes practice. Biographical or fiction writing often starts being used early in life and on a consistent basis, so it feels easier to write in this style. The need for apt technical writing skills, on the other hand, does not arise until much further in one's academic career. My technical writing still strikes me as flat, uninspired, and clunky; this new language is still too foreign for me to write comfortably and fluently in. So, I practice.

Do not let writing projects cause more consternation for you than any other type of homework assignment, and do not be frustrated or dismayed if you initially receive lower-than-expected grades on your scientific writing pieces. This is normal. To put this reality into perspective, expecting to be able to turn out a well-articulated document simply because you speak English is akin to expecting to be able to complete a marathon just because you can run. There is a great deal of training that goes into both, and with that training, both are very doable. Therefore, I will spend a fair amount of time in this chapter talking about the different kinds of writing and the best practices for good writing.

Believe me when I say that learning how to write well as a STEM practitioner will never count against you. Doing so is nothing but advantageous for you.

With this in mind, what are three ways in which you can improve your technical writing?

- Read good *and* bad writing. Stephen King reads seventy to eighty fiction books a year, and in his memoir *On Writing*, he is very specific about the importance of authors reading to improve their craft. His craft is fiction (the kind I am too tender minded to read), but his memoir taught me three of his habits that aid his writing: write every day, read every day, and edit out ten percent of your material before sending it on.

- Read technical papers. In this season of my life, my craft is technical writing, so I download technical papers (some good, some bad) and try to read one per day. I am slowly making my way through technical books on geomechanics, the compaction of argillaceous sediments, and discrete fracture networking. The idea is to read what you want to write.

- Write, write, write. In her writing memoir *Bird by Bird*, Anne Lamott suggests starting by writing three hundred words per day. Depending on the day, she may hit five hundred to one thousand words, but if it takes her all morning to get to three hundred, then so be it. Stephen King, meanwhile, writes a minimum of five hundred words per day, seven days a week. The liberation lies in the discipline. If you or I know we are going to sit down daily and write (and that it does not have to be great), then there is a win no matter what. Knowing ahead of time that material will be marked out, rearranged, and edited beyond recognition brings freedom.

- Have your writing critiqued. In college, my senior year roommate and I were not getting the grades we wanted on our papers: Bs, B+s, maybe an A-, but never the sought-after A. We swapped papers and gave them a read through to figure out the hangup. In my roommate's case, not a great deal of editing was needed, and the content was good; she just needed to finetune some of the grammar and structure. After hearing my comments, she asked me, alarmed, "Wasn't there anything you liked about it?" and of course there was! I just hadn't told her, because that hadn't been the focus of our session. After that, we did indeed earn our

As. My three-layer process for critiquing papers is: start with the overall aspects you appreciated; comment on how the writing can be improved; and, finally, end with specifics that you really liked and should be left alone. Find a classmate or a roommate, or work with someone in your university's writing center. Every time your writing is reviewed, you will see patterns for improvements you can make for the next assignment.[5]

How to Structure Your Technical Writing

I am going to date myself a bit here so that you understand in advance where these comments are coming from:

When my generation was (and older generations were) in school, we had to first create a study question—a hypothesis—that needed to be answered. Next, we wrote our outline on a sheet of paper. Then, we took notes from our reference materials, wrote them on little 3" x 5" notecards, and taped or pasted them to sheets of paper in an order that followed our outline. If we needed to change something around, we would cut a note out and paste it in a different location. Finally, when we had all our notes and sentences taped or pasted in order, we would take that holy mess (the rough draft) and type it up into a clean version to be reviewed and edited into the first draft. This is why we use the phrase "cut and paste" in a Word document or Google Doc today: people used to actually use scissors and tape (or paste) to construct their written documents back in the day (i.e., before the 1990s, when many of us started with Word Perfect). Now, with Word documents and Google Docs and a variety of options available to us on our laptops, we have a much more freeform way of writing.

In some ways, this allows for a more natural writing process, because it allows us to write as our thoughts come into our head; but it also means an increased level of discipline is needed if we are going to put together a tight,

---

[5] Do show your reviewer or editor respect and clean up your typos before you hand your written work over. Their time is valuable, and you do not want them to spend it fixing errors you can easily identify and fix yourself.

thoughtful, coherent paper. This means that when STEM professionals sit down to write their paper now, their approach is more like this:

They start with a blank document and put in the features they like (i.e., page numbers, a header and footer if needed, the font they like, and the correct font size, if there is one). They may or may not need to adjust the margins on their paper. Then, most have an idea that they are going to write about, and this would constitute their hypothesis. I encourage you to have this typed out at the top of your page. You need to know the direction you're going in, and you also need to know when to stop writing. (Whether you've answered the question or not determines whether you can stop writing.) Next, they put in any major and minor headings (as in, section titles) that they may or may not need. This is essentially their outline.

All reports for scientific and engineering research have the same layout: Introduction/Background, Methods, Results, Discussion, Conclusions, and Recommendations. You will also need a References section, so keep track of your sources and make note of those as appropriate. Most reports also require a title page, and some will require an Abstract. I recommend double checking the assignment requirements before you start out, as there's no point in doing unnecessary work unless you want to, and you certainly do not want to get to the end of the assignment only to discover you did not do what was asked and it all needs to be redone. If you've ever worked in construction, you'll know this rule as "measure twice, cuss once".

Now, this is the part that may not seem obvious: I know of virtually no one who starts at the very beginning of their paper and types it all the way through to the finish. Most people I know (STEM professional or not) insert their major section headings and then start putting their thoughts in under the appropriate one. Our brains don't always work chronologically, or even in an intuitive order; they tend to work more by association. One idea may trigger another, but it belongs in another section. So, absolutely feel free to work in this way, but remember that the more rigorous you are about keeping a tight paper (where only the content pertaining to your methods is in the Methods section, and the results are only in the Results section, and so on), the higher your grade will be.

If you are writing a term paper, use the same approach: start with a rough outline, use your major and minor headings, and then proceed to write your brilliant work in the appropriate section. I strongly recommend this system.

Again, you need to have the essay question at the beginning of the paper so you know whether you have built your case and answered it or not.

Once answered, you're done! Now, onto editing.

Editing

Every paper you write needs to be edited. Think of editing your paper as debugging your code. In the beginning, this feels overwhelming and exhausting (with no guarantee of success), but as you go through the process and develop a system, your editing will become smoother and faster.

Some papers will need more work than others, but no matter how strong of a writer you were in high school, trust me when I say you *need* to have your work reviewed.

There are roughly two camps of writer/editor: the first camp writes the paper all the way through and then goes back to the beginning and edits; the second camp edits as they go. Both methods are "correct", and the choice is simply down to the preferences of the author.

I personally tend to fall into the first camp. It works better for me to get all my thoughts down on paper and organize according to subject headings. Then, I can go through and start moving pieces around, tightening up the language, taking out contractions and putting in transition sentences, etc. The second way works for people who like to make sure they have every word perfect as they go, and if this is your method, go for it! Just be careful not to get wrapped around the axle over one sentence or word choice. If you find yourself stuck, leave it and move on. You can always come back to it.

Once you have gone through your paper yourself, hand it off to a friend who is willing to trade papers with you so that you can review each other's work, or take it to the writing center, if your campus happens to have one.

I made extensive use of the writing center when I was in Divinity School, and when I went back to school for engineering, I continued that practice, which had a twofold positive outcome:

1. My papers were always done early. Since I needed to have them completed with enough time to both be edited and given back to me to be finished up before the deadline, I was always ahead of schedule.
2. I started learning where my shortcomings were as a writer, and over time, I was able to edit those problems out before I handed off my papers to be reviewed. This meant that the hour a reviewer spent on my paper could be dedicated to increasingly sophisticated criticisms, and it also meant I became a better writer.

If you are able to use reviewers at a writing center on campus, or if you have access to professional reviewers through online tutoring sites, I strongly encourage you to take advantage of this. The ability to communicate is critical for conveying your ideas.

Do not hand off a "rough draft" version of your paper that you have not already edited and reviewed. If it is in shoddy shape, you will quickly frustrate the person reviewing your paper. Go through and remove sentence fragments, run-on sentences, and typos. Do not ask a reviewer to basically write your paper for you. This is lazy writing, and is not writing that deserves to be edited or reviewed.

How do you self-edit? Here are a few quick fixes you can make before sending your "first draft" version of your paper for editing/reviewing:
Confirm the page length, font, size, and margin requirements for your assignment.

- Clear out all contractions and colloquialisms.
- Use cliches sparingly.
- Use nine-dollar college words sparingly.
- Double check spelling (especially homonyms that spellcheck will miss).
- Make your Oxford comma usage consistent.
- Shorten run-on sentences, or unpack extremely dense sentences.
- Double check your references and ensure they line up.

- Go back and scan your document again for red lines. There is no excuse for missing these misspelled words.
- One space after the period at the end of a sentence, only and always.
- Check the spelling of every word in your title. Seriously. This one gets everyone at some point.
- Ask: do all your verb tenses agree (i.e., is the paper consistently in the past tense or the present tense)?
- Ask: is there a clear statement of purpose in your introductory paragraph?
- Ask: did you follow your statement of purpose up with three to five very brief descriptions about how you intend to support this contention?
- Ask: which "person" will you write in, first or third? Be consistent.
- Ask: do all your transition words and phrases have commas after them?
- Ask: are your supporting paragraphs fairly and evenly weighted?
- Ask: do you have a paragraph exceeding the length of one page? If so, chances are you need to shorten it.

The other mistake people can make (and I have done this) is adding new material to their paper every time they get it back from their editor. Instead of doing this, have all your substantive work done and in the paper before you hand it off to be edited or reviewed.[6]

My last piece of advice is, be available to review other writers' work. If people are willing to review your work, then be willing to review theirs. The further along you go in your career, the harder it will be to avail yourself of someone's editing abilities, so take advantage where you can and help others when your time comes. This will help you in your own writing as you start to see common mistakes in others' writing. Take the time to eliminate those mistakes in your own work before handing it off.

---

[6] This pertains more to a thesis-level paper or a peer reviewed paper, but even in standard papers and homework assignments, it is a good idea to practice not adding brand-new material after your work has been reviewed/edited. Again, this will frustrate your reviewer or editor, and you will likely lose their largesse for future work.

When to Ask for Help

You will, at some point in your academic career (more than likely in college), come across a subject, concept, or homework problem that is just inscrutable. Fear not, even if you have never experienced this before. We have all been there. My recommendations for tackling college-level homework problems are:

1. Read the problem or concept through two or three times—out loud, if you need to.
2. If it is a homework problem, write out all the information, including a brief problem description, the known variables, the unknown variables, and equations you think you will need—and, of course, draw a free body diagram, or sketch a diagram of what you think is being asked. Sometimes, seeing the problem laid out clearly will help you to see the solution. Occasionally, textbooks get it wrong and your confusion is justified.
3. Take a couple of good swipes at it, following any example problems that you might have in your notes or textbook.

If you still are not able to solve the problem, don't get frustrated. Set it aside and move onto the next one. Presumably, you have found a friend or two in class that you can review homework with, and at this point, I recommend doing so. If, however, you haven't had the chance to make these friends yet, or are not able to (perhaps you are taking an online course, or are working full-time while squeezing classes in at night, or are on your lunch hour), seek out a tutor or go see your professor. Many professors welcome visits from their students. Confirm your professors' office hours and sign up for them, if that is needed. Make sure you have reviewed your textbook before you visit, though. There is nothing more aggravating to a professor or a TA than a student showing up complaining that they cannot figure out a homework assignment, only to discover they never actually opened their textbook.

Some problems are straightforward, and you will be able to solve them by having your colleague, tutor, or professor walk you through it. Other times, concepts are more complicated and will take longer to work through. Again, this is normal. You are learning a great deal of information all at once, and I suspect much of it you have not heard of before. You are going to take several passes at it before it really sinks in. Again, this is normal. Show yourself some grace and just keep practicing. It *will* start to make sense at some point.

Productivity versus Effectiveness

There is a difference between being productive and being effective, and this was most clearly shown to me in Benjamin Franklin's autobiography.

As far as human beings go, B. Franklin, as we have discussed, managed staggering productivity and effectiveness on several fronts throughout his lifetime. Toward the middle of his autobiography, he listed out his day hour by hour, top to bottom. From that, I know that at the beginning of each day, he asked himself, "What good am I going to do today?" At the end, he asked, "What good did I do today?"

The title of this section ("Productivity versus Effectiveness") implies that these two things are mutually exclusive, but in fact, productivity *contributes* to effectiveness. To define both, though, being productive means producing much, while being effective means being *successful* in producing an *intended result*. Thus, I modify B. Franklin's question to ask myself, "What effective movement am I going to (or did I) make on my research today?"

Rather interesting questions, my friends.

"Effectiveness", in the world of STEM studies, means contributing to the body of knowledge: advancing science, innovating new technologies, solving intractable engineering problems, or improving civilization. We need productivity in order to be effective, but it is your effectiveness that upends conventional thinking, gives you purpose, and changes the world. Effectiveness moves "the thing" forward, whatever your "thing" may be. Productivity requires little critical thought or creative understanding, while effectiveness demands critical analysis and creative problem solving. Effective contribution takes time.

Marching through a to-do list of short, simple tasks gives me great satisfaction. Here, I am being productive. If anyone asks me what I accomplished on such a day, I can rattle off fifty finished items. Demonstrating effectiveness in one's day, on the other hand, risks the appearance of dilly-dallying or incremental (if any) movement forward. To the untrained eye, collecting rock samples looks like an excuse to go hiking on a sunny day (even if it will ultimately contribute to an impressive body of work). When a well-intended loved one asks me what I got done on an effective day, my thin answer of, "I tracked down the platens I needed and took measurements so I can have new ones machined," leaves me dispirited. Why, oh, why do I not have fifty finished tasks?

Research in a new area, while ultimately an effective endeavor, feels much like slowly groping one's way forward in the dark, in a foreign environment, with few recognizable landmarks. The process is slow. I long for quick turnaround and sharp, decisive steps. Nevertheless, I have had fast-paced jobs that capitalized on high productivity and low intellectual challenge, and let me tell you, my interest in these tasks wanes very quickly and leaves me to wander off in search of something interesting; something challenging; something that needs figuring out.

Linked to productivity versus effectiveness, another dance exists in this intellectual space: perfectionism versus excellence. Perfectionism versus excellence is learning every conceivable detail about every conceivable process/instrument/shale formation, versus learning what I need to know in order to understand the work. Perfectionism versus excellence is reading just one more paper or text in case I missed something, versus drawing the finish line and calling it done. Aim for effectiveness over productivity and excellence over perfectionism.

The Point of Diminishing Returns

There are two different points of diminishing returns when it comes to your homework and study efforts. First, let's talk about homework and grades.

This advice was given to me by an aeronautical engineer when I entered my engineering program. She made it very clear that while studying hard and working toward good grades is important, there is a point when you need to hit the brakes. If, for example, a homework assignment takes you forty hours to earn a B, while an A would take four hundred hours, take the B and move on. The extra amount of effort needed to earn the A simply is not worth it when there is so much other work to be done and so many other ventures to be pursued.

This advice has proven invaluable.

First, it gives us permission to know when to be done. Similarly, it also helps us to establish what exactly the end goal is. If the end goal is to learn the material rather than to attain a certain grade (which I, for one, believe it should be), your experience will be much more satisfying and meaningful.

Second, it shows the reality of going after some of these high-level goals. Sometimes, they simply are not worth the effort required to achieve them. My goal was never a 4.0 every semester: I always endeavored to learn the material and earn good grades, but my focus was not on the GPA itself.

For some of you, the GPA or grade certainly will be the focus. Some of you reading this book will want to be the valedictorian, or to graduate *magna cum laude*. I have great respect for those accolades, and applaud your efforts. If that is your goal, however, know that (with rare exception) you will need to be dedicated to those pursuits one hundred percent of the time. You will also need to forgo much of the rest of the college experience in order to achieve it. I, personally, was interested in developing my social networks, conducting research, and working to offset some of my student loans, so for me, that GPA simply was not as important. This will be an individual decision, and I advise you to take a bit of time to think about what you want to do and the sacrifices required to do it.

Now to talk about the second point of diminishing returns. When you have been working so long and so hard, there is (as a mentor of mine used to say) "keeping your nose to the grindstone, and then there is deforming your face". If you have had your nose to the grindstone for so long and worked so hard that you have worn your nose right off, it is time to take a break. Go do something else.[7]

That said, the cumulative effects of a semester can unfortunately still add up even when you are being diligent about getting good sleep and taking your rest days. If you have gotten to the point where your neck muscles are seizing up, you've burst a blood vessel in your eyeball from staring at a screen, or you really do not recognize the humans around you, then it is time to put the books away and go for a walk. In fact, you may need to take a mental health day (see Chapter 11).

In the end, you need to work hard to learn what you need to learn and earn the grades you need to earn, but you also need to use this time to learn when to stop and move onto the next project.

Save Your Work

I recommend saving your work in an organized and accessible fashion throughout your college career. When studying STEM disciplines, you may find that you ultimately want to go on to get a professional license, and when doing this, referring to past homework assignments (particularly in your specialized field) will be helpful.

In the engineering world, the general initial exam we all take is the FE (the Fundamentals of Engineering exam). This exam covers all the basic topics that are required for engineering courses. There are some sub-specializations or fields for civil, chemical, electrical and computer, environmental, industrial and systems, mechanical, and other engineering disciplines, but if you are not in one of the designated specializations, I recommend just taking the "other disciplines" exam.

---

[7] Pro Tip: If you are taking your weekly day off, facial deformation is less likely to happen.

The next exam is the Principles and Practice of Engineering exam for your professional engineering license. This exam is more specialized for the varieties of engineering available. The same is true for the sciences—the professional geologists (PGs), for example.

Hold onto your assignments and look to the "Paper or Digital?" section (next) for guidance on how to decide whether you should save them in paper or digital format. Some of your larger projects you may also want to hold onto as examples for a portfolio, if you are considering graduate school. If, alternatively, you have no intention of going back to school after you graduate, then recycling your work is perfectly valid and a way to clear off your dusty shelves in the basement.

Homework, Quizzes, and Exams

You will not need to keep your homework, quizzes, and exams indefinitely, but I do recommend storing them for the duration of your college career. At the end of each semester (provided you maintain your course binders, as will be discussed in Chapter 6), you should be able to assemble each semester into one large four-inch binder. Label it with the semester and year and then tuck it away on a shelf somewhere out of the way.

I recommend including in these binders the syllabi for all your classes. These have, oddly, proven helpful for a variety of reasons, including requesting transfer credit or justifying an independent study, or in graduate school, when I was trying to illustrate why I was eligible to take an advanced class even if a prerequisite ("prereq") had not been taken at that university.

You may want to hold onto these binders if you intend to pursue professional licensure so you can practice example problems for your exams. Keeping your homework binders tidy and organized and on a shelf out of the way will end up saving you headaches later if you need to track down this information.

Paper or Digital?

Whether you should save your work in physical or digital format depends on the intent behind you saving your documents.

I do not believe that you need to go through all the trouble of scanning each homework assignment, quiz, or exam; those are fine to stay in paper format in a binder. Papers and projects are sometimes constructed and delivered in digital format, so you may want to print out documents that have particular meaning for you.

My first bachelor's degree was in the late nineties, which means that all my work was done on a Compaq 386. I retained none of those digital files, and even if I still had that computer, I would not be able to access or read them at this point. My point here is, the challenge with digital methods of saving work is that they work for right now, but twenty years hence, our current technology will be obsolete. For this reason, the papers that I saved from school are in paper format. There is, however, an appropriate purpose for saving in a digital format: creating backups. See the next section.

Backups

There are many ways in which you can back up your work, and these methods range from cheap and kludgy to expensive and thorough. One thing is for sure: there is no excuse to *not* back up your work. A quick search online will show you just how many master's and PhD students have lost the data for their research (or thesis or dissertation) and are desperate to get it back. Laptops can be stolen from coffee shops, backpacks can be left on the bus with thumb drives in them... you name it. This *actually happens* to real people, in real life, and you will do well to remember that.

When I was working on my master's thesis, I had my data backed up everywhere, because losing it would have meant starting all the way back at the beginning. I could always reconstruct my writing, but the data could not be replicated. Once it's gone, it's gone.

If you lose your data, or your hard drive pukes, or your materials are stolen, the very first question you will be asked is, "Did you back up your files?" and you do not want to be the STEM student who stares back sheepishly and says, "No." With the abundance of mechanisms we have available to save our work, there is no excuse to *not* do so.

When deciding what backup method to use, you will want to consider your threshold for pain when it comes to reconstructing work and wasting time. It's a bit like car insurance: if you want a really low premium, you may end up with a really high deductible, but if you don't have that kind of cash on hand, you may end up paying a higher monthly premium so you have a low deductible, should an accident occur.

The same goes for backing up your work. The more time and effort you put into saving your work and the greater the variety of places in which you save, the less work you will have to do reconstructing it, should something happen. There are certain rules of the universe that I just accept and do not question, and this is one of them. If you pay for car insurance, you probably won't need it. The more you save your work, the less you will need to actually use these means of saving your work.

Let's look at the cheap and kludgy versions first. The easiest way to back up your work is to simply email your document from one email account to another. Period. Presumably, you have a personal and a school email account. Just by emailing your document from one account to the other, you have now effectively saved it in two places.[8] This also means that every time you email your report or master's thesis (or doctoral dissertation) to a colleague to review, you are effectively saving a version of it. This can be confusing, so make note in the title (Report V1, Report V2, Report V3, etc.). If you continually save over the prior document, you may lose changes that you want to go back and reincorporate, or reconstruct.

The next easiest way is to use an external drive. In fact, I recommend saving on two or three thumb drives if you have them available. You can also use cloud sources like Dropbox. These methods are suitable for keeping

---

[8] Obviously do not delete the emails!

either one or two files safe, or for backing up your entire hard drive.

Storage is cheap, so for these important and larger documents, save liberally. My advisor used to say, "Jesus saves. So should you." For backing up your entire hard drive, certainly use an external drive that you keep in a safe place. Author Steven Pressfield, in his book *The War of Art*, explains that he saves his work at the end of every day on a disk, takes the disk outside, and locks it in his pickup's glovebox, in case his home office burns down.

I do not quite go to that level, but I understand the inclination.

I have my computer set to auto-backup so I do not have to worry about it. For documents that I am particularly concerned about, this method works. For pieces that are not replicable or would take an inordinate amount of time to reconstruct, I also periodically save those by emailing the text file to a couple of email accounts, in addition to using my hard drive.

Again, there is no excuse to *not* save your work, and the consequences of it can be devastating.

Recommended Reading

- *On Writing: A Memoir of the Craft* by Stephen King.
- *Bird by Bird* by Anne Lamott.
- *Mindset: The New Psychology of Success* by Carol S. Dweck.
- *The War of Art* by Steven Pressfield.

Strategies for Success: Homework

- Set up and turn in homework that represents your level of professionalism.
- Know that all learning is governed by reps.
- If you need to memorize, write the material on large notecards and review them while on a walk.

- Advanced STEM courses all build on the introductory courses you take. Take the time to learn the material early and you will save yourself effort in the future.
- Always edit your work—even better if you have someone read it before you turn in your final draft.
- Save your work often and in different locations.

5

Procrastination

Done is better than perfect.
—SHERYL SANDBERG

So get it done.
—SCYLLER'S MOM

A MY POEHLER WRITES IN HER book *Yes, Please,* "Talking about the thing is not the thing. Doing the thing is the thing." To this, I would add that blogging about "the thing", thinking about "the thing", or napping before starting "the thing", while all important, are also not "the thing".

Why do we so often find ourselves doing anything but "the thing"? Three reasons come to mind: no time, no idea what step comes next, or no interest. Two of these "reasons" are a fallacy: you *do* have the time and the skill, indisputably.

On my morning walk today, I read a quote on my favorite quote-filled marquis: "The skill to do comes in doing." In his book *How to Stop Worrying and Start Living,* Dale Carnegie argues that taking action (no matter how small that action is) is enough to stave off the anxiety of... well, of (not) doing the thing.

I love epic, sweeping projects. I fantasize about a bold finish with a considerable impact. I relish creating the plan for starting "the thing". But seeing "the thing" through after the enthusiasm passes is when discipline is needed. To get through procrastination and complete major projects ("the thing"), we need to be doing something that requires daily effort.

Cue the dailiness of discipline.

Enthusiasm is the catalyst for an epic, sweeping project. Discipline allows me to tap the project along even when my enthusiasm wanes. If I stop showing up daily, that same enthusiasm (which is drawn to shiny objects) will threaten to pin me to the couch and strongly urge me to finish reading a book on [insert random subject matter here] instead of doing "the thing". Or suddenly, the kitchen will *need* to be cleaned from top to bottom. Or my blog will *need* to be updated and reorganized. Or I *must* catch up with the friends I have neglected for a semester. Or my plants will need watering. Or I'll suddenly realize my laundry is piling up. Hey, when did I last knit a row? When did I last walk my dog?

My quotidian quagmire.

And then there is the procrastination that often comes with possibility. Shiny objects tempt many of us—and by "shiny objects", I mean the new and interesting projects filling our imaginations. "As a matter of fact, I *do* have an idea for an energy resources course I could develop and teach next year!" "Why, yes, I *would* love to write a study guide for [insert random subject matter here] this summer!" I message friends as fellow collaborators, but I do not tell them that I am yet to finish "the thing" because then they will all know that I am supposed to be focusing my efforts on "the thing" and that my shiny new idea is mere procrastination.

Suddenly, "the thing" is boring and I want it to go away.

When projects get too big, cumbersome, or long, it feels infinitely easier to start and finish a small project, even if that small project has nothing to do with "the thing" (remember that discussion we had about productivity versus effectiveness earlier?). For example, while working on my dissertation, it occurred to me that I had been meaning to read *Anna*

Karenina for more than a decade. Thus, I promptly located a copy and read it from cover to cover (and yes, I should have been working on my dissertation).

Really, Scyller? *Tolstoy?*

I take great comfort in knowing that artists, writers, and researchers all struggle with projects grinding to a standstill. Under these suffocating conditions, creativity gasps for life. Not all is lost, however: this is where the daily discipline of showing up breathes in new energy and propels projects to the finish line. My daily "showing up" goes on my calendar just like a non-negotiable appointment. I know that putting in a certain number of hours a day six days a week will get me to the end, and that the end is in sight.

During my master's program, I wrote, "Done is better than perfect. So get it done," on my office whiteboard. Do "the thing". Finish "the thing". Greater adventures await.

Why Do We Procrastinate?

Well, for a few good reasons.

- "Beyond writing my name at the top of the page, I have no idea how to get started." This happened to me during a couple of major engineering projects. The remedy? Set your ego aside and ask for help. Open a conversation with, "I have given this a good think and really have no idea how to get started," and see where the discussion takes you.
- "I have started, but now I am stuck." Typically, this means I have a good idea for the project in question, but I do not have anything new to add. In other words, my depth of understanding is not plentiful enough for insightful writing to abound. The remedy? Head back to the library, pillage the bibliography and references from published works, and start reading and taking notes. If you've run out of material to work with, it's time to add more material.
- "I physically cannot do it (because I am falling over from lack of sleep, or I cannot sit still, or my stomach will not stop begging for necessary

sustenance).” The remedy? A power nap, quick workout, or meal—whichever applies.

There is also one bad reason for why we procrastinate: we just do not want to do it. The remedy? In the words of a former roommate: suck it up, cupcake. Really, just set your timer and go. You will never want to do something more than you do right now. Putting it off will not make the task more attractive.[9]

Productive Procrastination

As odd as it sounds, there are a couple of benefits to “deliberate procrastination”. Once the starter material is in my brain, it has the time to sit and ferment into something really interesting. I do not start writing or attacking the problem until I am closer to the deadline precisely because the interim rumination phase makes such a big difference. Applying the mechanics of writing or problem solving is less of a challenge than developing the “big idea”, and the latter often takes time.

It also gives me the time to run my idea past people “in the know” and elicit feedback. Discussing and refining ideas can make them stronger, provided you follow through and put the upgraded ideas into your paper or projects.

Gulp List

I have developed my Thursday Morning Gulp List (TMGL) methodology over the course of several years. What is my TMGL, I hear you ask? The TMGL is comprised of the prior seven days’ worth of one-off administrative

---

[9] Note: If you find that you are in a degree program (or a job) where this is your standard response rather than an occasional episode, there is a good chance you are a bad fit. The remedy? Find a new adventure.

items that need to be completed. You know the ones; they probably trigger an inexplicable dread in your gut if you don't have a reliable system for them.

Here's an example of a Gulp List:

- Thank-you card, scholarship committee.
- Thank-you card, care package.
- Purchase airline tickets, girlfriend's wedding in June.
- Secure hotel room, girlfriend's wedding in June.
- Find summer living space.
- Respond to emails (Ms. X, Mr. Y, Ms. Z).
- Get airline miles applied to account.
- Get bank fee removed.
- Submit scholarship application.
- Plan summer internship route.

The TMGL does not include daily or weekly routine items (household chores, paying bills, workouts, and so on); only items which need to be scheduled in if they're ever going to be done.

The reason this list works for me is, I spend three to four hours of my mornings on creative/critical thinking projects (mainly my research and attendant writing), and it is while I am working on these things that any "Gulp List"-esque items needing my attention will occur to me. Since they do not need to be done that very minute, these items are written down on my Thursday morning calendar (my TMGL). This clears the space in my head for what really matters—my "One Thing" Keller and Papasan talk about in their book *The One Thing: The Surprisingly Simple Truth Behind Extraordinary Results*—and allows my brain to rest easy knowing these items will not be forgotten and will get done.

I chose Thursday morning for two reasons. First, come Monday morning, I am excited to get after my interesting, brain-intensive work (not soul-sucking administrative tasks). This energy carries me solidly through to Wednesday. By Thursday morning, my brain needs a change of pace. Second, attacking my Gulp List on Thursday morning means that if I am waiting on a return call, there is still a day and a half left of the business week left to get the task handled, meaning I don't have it kick over into the next week. Waiting until Friday afternoon to get after my Gulp List meant several items kept carrying over, denying me the satisfaction of a completed list at the week's end (and any claim to a "productive" day!).

Note that there must be a cutoff for your list. Mine is Wednesday evening. Without a weekly cutoff, items just keep getting added, and the list is never done. Having a finish line is *imperative*.

Use these strategies for success when writing your Gulp List, and you'll be gold:

- Only include one-off tasks.
- Only include items that are important but not in need of immediate attention.
- Have a list cutoff. If this is Wednesday evening, any items that are added after that point are to be completed the following Thursday.

Multitasking

Keller and Papasan make a very compelling argument regarding why multitasking does not work for important tasks, and I agree with them wholeheartedly.

There are a couple of occasions where multitasking is appropriate and will help you get through some of the more tedious aspects of your day. For example, if you're watching television, this is when you should be folding your laundry. If you want to listen to an audiobook, do so while you're out for a walk or run or while you are cleaning up the kitchen. If you are memorizing formulas and have them written on your notecards, keep those in front of you while you are tidying the house. You can double-dip your time with one tedious task and one fun item when you do not have to have your undivided attention on something, and this will allow you to enjoy the best of both worlds.

When you are studying, however, I strongly recommend that you simply focus on the task at hand. Do not try to respond to emails or text messages while you are doing your calculus problems, or to solve physics problems while watching the game on TV. That is a waste of time. If you want to watch the game, mark that as personal time and simply go and enjoy it. Multitasking has its place, but not when you are doing the heavy-duty intellectual lifting that you need to do to learn your subjects.

Now, if you will excuse me, I am off to watch *Star Trek* and fold my laundry.

Stop Procrastinating

Let's consider the procrastination of tasks that should be done and for which there will be no benefit to putting off.

As I said before, you will not want to do any task more than you do right now, and so not starting it and letting it weigh on you makes approaching the task even worse. In fact, this can keep you up at night (see the "Physical Health" section in Chapter 11).

Just as we have different reasons behind why we procrastinate, we need to have different methods for stopping our procrastination.

First, let's take the situation where you just can't seem to get started. This is where having a timer is quite handy, whether on your phone, a kitchen timer... any sort will do. When I need to sit down and crank out a formal letter or proposal, or any piece of written work that I know I can realistically get done in a decent amount of time, I set my timer for ninety minutes. I never set it for longer than that, as that's usually enough to get me to sit down and write out whatever piece needs to be completed.

There are times when I need to start a homework assignment that I'm fairly certain will take longer than that, but I still need to at least get it started. So, I will set my timer for fifty-nine minutes (I like fifty-nine minutes because you know that it is not quite an hour). During this time, I know I can sit down and make a good go of it. It probably won't get ticked off the list completely, but I will have made enough progress to have cut down my future resistance to doing it by a significant fraction.

And then there are the occasions when you are just dragging yourself to the start line and fifty-nine minutes is too long, never mind ninety. In this situation, twenty-five minutes is doable. This method it is also called the Pomodoro Method, and there are numerous books, articles, and videos on the subject. In brief, it is a mechanism to just get you started by giving you an achievable end point.

The next method for conquering procrastination is to find three actions you can take to move your project forward that day. This method works for longer-term projects, like a semester-long paper or a senior project. I even used it for my dissertation. There were days when I had no problem moving along working on my dissertation, doing the research and typing up my notes. There were also some days when I simply did not have it in me, but I still knew my commitment to myself was to move my dissertation forward every single day. Well, except for Sundays.

This method is fun because you start learning how to break your project down into the tiniest possible components. Every effort counts, so you just need to come up with three, and then you can count your project as having been moved forward for the day. It could be something as straightforward as sending an email to set up a necessary meeting, typing up a source for your bibliography, or locating and captioning a figure. The fun part is, if you can come up with a task (no matter how small it is) and you mark it off your list, then it counts.

I recommend this because you will always feel better, no matter how small the step forward, if you mark three things off your project's to-do list every day. If you have taken calculus, you'll know it doesn't matter how infinitely small the parts are! When you integrate them all together over the course of the semester or degree program, you accomplish something quite remarkable.

The last method is quite powerful, and is one that gets you up and off the sofa, out of bed, and onto the next task: Mel Robbins wrote about the five-second rule, and it works wonders. You count down from five to one, and at one, you begin. Do you have something you are supposed to be working on right now? Your "one thing"? "The thing"? Start counting down from five... and go do it! Come back and finish this book later!

Recommended Reading

- *The 5 Second Rule: Transform Your Life, Work, and Confidence with Everyday Courage* by Mel Robbins.
- *The One Thing: The Surprisingly Simple Truth Behind Extraordinary Results* by Gary Keller, with Jay Papasan.
- *How to Stop Worrying and Start Living: Time Tested Methods for Conquering Worry* by Dale Carnegie.
- *Daily Rituals: How Artists Work* by Mason Curry.
- *Daily Rituals: How Women Work* by Mason Curry.

Strategies for Success: Procrastination

- The desire to procrastinate arises in everyone. Discipline will see you through more than motivation or enthusiasm will.
- Use a timer to give yourself a start and finish time to your work (ninety, fifty-nine, or twenty-five minutes).
- Use productive procrastination judiciously.
- Multitask tedious work with something fun, like an audiobook or episode of your favorite show.
- Take a nap, have a meal, or do a workout if you cannot focus.
- Give yourself one regular time a week to complete your Gulp List.
- You will never want to do a task more than you do right now, so go after it.

6

Organization

For every minute spent in organizing, an hour is earned.
—BENJAMIN FRANKLIN

T HERE ARE MORE BOOKS WRITTEN on organization than just about any other topic on the planet right now. This chapter is solely dedicated to the organizational habits that will make your college life more effective and productive. The ultimate goal is that you will carry these habits with you into your next phase in life, be it in work, the military, or graduate school.

There are two extremes on the organization continuum: you can be so completely disorganized that you are spinning in circles, accomplishing nothing, or you can spend so much time focused on your organization that you never actually get around to accomplishing anything. Neither of these extremes are desirable; you want a middle ground.

Remember that your organization is there to help you manage your day more effectively, not to stress you out about whether you have maintained part of your organizational system to a T. Accordingly, when organizing, the first question you have to ask yourself is, "Are these practices helping me be productive, effective, or both?"

How we arrange our classes, days, weeks, months, and semesters will have a tremendous impact on how much we learn and accomplish. We only have so much time and energy, so how we choose to spend this time and energy is critical. A few minutes spent organizing at the front end will give you hours of time to be spent on better, more productive endeavors later on, whether that is your homework or a hike!

Tools of the Academic Trade

I will proceed with the expectation that you already know you need a laptop, non-programmable calculator, crockpot, and notepaper as you head into college life. Beside these staples, the following items are at the top of my must-have list when it comes to making academic life a little more efficient and a lot more fun:

- Earplugs. I wear them during exams (who needs to hear sonorous mucus gurgling or the panicked rattling of other students when you're trying to keep a lid on your nerves?), when I need deep concentration to problem solve, and when I sleep. These are available at drugstores and in the hunting department of sporting goods stores.

- Soft-sided three-ring binders. Buy five or six (one per class for a semester) and reuse every semester thereafter. These are lighter and fit into your backpack much easier than hard-sided binders.

- Three-hole punch with 13/32 diameter, not the typical 9/32 diameter hole. A luxury item, maybe, but absolutely worth it if you find you read and store reams of notes or technical papers in binders. I will never go back to 9/32. Heed my advice, and you will flip through pages with ease and whimsy!

- Colored pens. I used to use highlighters, but reading and notetaking with colored pens is much more efficient and allows me to differentiate topics by color within a given paper. I write questions, comments, thoughts, and definitions to myself in the margins—something I cannot do with a highlighter.

- A mechanical pencil (with spare lead). Get a real mechanical pencil and do *not* loan it to anyone. I use a 0.5mm HB. The graphite is dark enough to be seen on engineering paper, but not so dark that when I try to erase it, faint traces remain. Oh, and never go into an exam without spare lead.
- A separate eraser. As much as I love my mechanical pencil, I have yet to find one with a sufficient eraser. Buy a separate one.
- A reusable water bottle. Dehydration leads to brain fog. In school or not, our bodies and brains need three to four liters (one gallon) of water a day, minimum (more if you are an athlete, and even more if you drink copious amounts of coffee).
- Glass food containers. Please do not ever reheat food in plastic containers. Purchase a few glass containers in which you can store your crockpot leftovers. Glass containers work beautifully for both freezing or reheating in an oven or microwave.
- A calendar. For community scheduling purposes, I use Google Calendar. Fellow students and professors use this to request meetings around recurring classes and one-off appointments. My weekly planner, however, contains the specifics of what I plan to get done and when every day. I cannot imagine anyone wants to know when I plan to grade labs, write an essay, or clean my bathroom, so this is for my eyes only.
- An external backup. I use a simple external hard drive. You do not need anything fancy, but you need to back up your work. Ask anyone who did not... and lost it.

For some "bonus materials" inspiration: I carry a small zippered bag with my pens and pencils and extra pencil lead, erasers, and cough drops. The purpose of these is to help a fellow student out. If you always have an abundance of lead, you can literally save someone on test day. Same with erasers. Or you may have someone behind you trying desperately not to cough and failing at it spectacularly. Discreetly give them a cough drop. Someone showed me a kindness once with pencil lead right before an exam, and I never forgot it. These simple gestures can save the day for a stranger, and possibly introduce you to a new friend.

Binders

Class binders are either a hot mess or the best possible way to keep your classwork in order.

Hard-backed binders are fine; they do make for a fine impromptu writing surface. That said, soft-sided are my preference, given that they are lighter and more flexible (and can therefore fit into my backpack or side bag easier). Plus, soft-sided binders last a long time, so after the initial investment, I do not need to purchase them again for subsequent semesters.

I use heavy-duty dividers, with a pocket in the first divider. Assignments that are due go in the very first pocket as soon as they are completed so they are ready to be turned in at the beginning of class. This way, I am not scrambling to find them, and my homework stays neat/presentable.

I use one binder per class and divide each binder into five sections: syllabus (and only the syllabus), notes from class, returned/graded homework, returned/graded quizzes and exams, and auxiliary handouts.

- The syllabus is your contract with the teacher or professor. Typically, it gives vital information concerning class (and/or lab) meeting times and locations, your instructor's contact information, required and recommended reading, topics to be covered (i.e., what you are responsible for knowing by semester's end), homework deadlines, and exam dates. In graduate school, I did not read my syllabus carefully enough and turned in a take-home exam one day late. Unbeknownst to me, the professor had written in his syllabus, "LATE EXAMS WILL RESULT IN AN F FOR THE COURSE." So, he could have failed me. He instead dropped my final grade one whole letter grade. While this was better than an F, to have worked the entire semester for a solid A, only to end up with a B because I did not read the syllabus, was devastating, and it still aggravates me—and that was two decades ago!

- Notes from class are those which you handwrite while attending class. Put the date up in the right-hand corner (even if you are left-handed) and write all notes in chronological order. The instructor may give you daily handouts for notes, or they may expect you to take notes on your own paper. Either way, writing down what you hear will reinforce the subject matter being covered, and that makes studying for test day *much* easier.

- Returned/graded homework is exactly that. This is a solid bank of information from which to study. Most instructors do not have the time or inclination to make up brand-new, never-seen-before test questions for exams, and in many cases, this is not even possible, so you'll develop a solid understanding of what your instructor will test on during exams by studying what you got right and what needed correcting in your homework.
- The same story applies to returned/graded quizzes and exams. Use returned quizzes to study for the upcoming exam, and use returned exams to study for the final exam (if your final exam is cumulative).
- Auxiliary handouts are those which your instructor wants you to have for reference. They may or may not be used for exams. They may be necessary for you to complete homework assignments or group projects. If your instructor hands them out, there is a reason for it. Hang onto them.

Some students have tried keeping all their classes in one binder, and I find that the sheer volume of notes and returned assignments is much too cumbersome for this. Plus, if that one binder is lost, then all your notes and former assignments for all your classes are gone. Suffice to say I do not recommend this method.

Another variation of binder use that I have seen involves putting every piece of paper in your binder in chronological order. I have tried this on several occasions, and it just does not work well. Notes that should stay together end up split apart by handed-back material and auxiliary handouts; returned homework gets lost in the shuffle; example problem sets that could be helpful for study get bogged down by layers of notes. Instead, keep each section divided, and you will always know where to look.

You should not need to add additional sections to your binder. Add much more than what I've suggested, and the system gets too complicated to keep up with. Have fewer sections, and the binder becomes disorganized and inutile.

There may be classes where you have a staggering number of class notes or handouts. If this is the case, only keep the current material in your binder. If your semester is separated by four exams, then after you complete Exam 1, put all material in a large four-inch binder which can stay at home. Start your course binder with the material that will be covered on Exam 2. My advanced engineering classes were like this, and I ended up filling whole four-inch binders at home over the course of the semester. Each exam was a standalone, so I did not need to keep all my materials with me daily. When it became time to study for the final exam, the entire semester's worth of work was already in place, in order, and ready to go.

Your Schedule and Managing Your Time

Managing your calendar will be a critical skill for the rest of your academic and professional life. There are no shortages on how you can spend your time, for good or for bad, so we will spend a fair amount of time discussing your time in this chapter. By the end of this section, you will have learned how to:

- Decide what types of calendars or planners you want to use.
- Estimate the amount of time needed for homework and projects.
- Note deadlines in your planner.
- Decide when your day starts (and why this matters).
- Understand and make use of the need for study appointments.
- Understand and make use of (equally crucial) personal time.
- Formulate and follow a tailored morning and evening routine.

A significant portion of my (steep) learning curve upon returning to school constituted learning how to accurately estimate the amount of time needed for projects/homework/exam study, and how to locate that necessary time on my calendar. Time will get away from you before you blink if you do not budget it jealously with a calendar. If a financial budget tells our money where to go (so we don't wonder where it went), a to-do calendar does the same thing with our time.

As mentioned previously, I use both an electronic calendar and a paper version to manage my time. I note regular meetings, class times, or appointments (including my dedicated planned personal time) on Google Calendar, which automatically syncs to all my devices on which I am logged in on Google (you do not need to use this particular calendaring system, just make sure yours is reliable and easily accessible). I also keep a little planner that I fill in with an actual pen which displays seven days at a time. On each day, I have two lists: "Must Get Done" and "Would Be Great to Get Done". My lists include everything from workouts to work projects to homework.

No one else needs to see this, which is why I do not put these lists on Google Calendar. You do, however, need to block out time on your electronic calendar for completing your to-do lists. Others will want to fill up your time for you. Do not let them. Be aggressive and deliberate with *your* time.

Estimating how long homework assignments will take and slotting that into the blocks of time between classes can be tricky. To do this, you will use a combination of your personal experience and this rule of thumb: for every hour in class, expect three hours of homework. Also, if your homework generally takes ninety minutes to complete, then do not plan to do it in sixty minutes. Obviously. I am an optimist, but even I know you cannot cut one third of the necessary time out by "just working faster". If you tend to work faster, you can shorten the amount of time you put aside by a small margin, but not much. If you tend to need a little extra time because a subject is eluding you, then book in additional time to learn the material.

To estimate project (not homework) length, at a minimum, increase your estimate by fifty percent, or, more generously, by one hundred percent. This is your factor of safety. Human beings (including this one!) are notoriously bad at estimating the amount of time it will take to finish a long project, so accept this fact and account for it. For example, if you have a project you will be working on over the course of the semester and, after reading through the requirements, you estimate it should take you fifteen hours (an hour a week for the duration of the semester), plan to spend ninety minutes to two hours a week working on it.

The good thing about this approach is that it sets realistic boundaries around what you can accomplish in a given day. We can all tell ourselves that we will get up early, pound out fifteen hours from 6AM to 9PM, and just knock that semester project out in one Saturday, but first, that will never happen, and second, you know from experience that your initial estimate is always wildly shorter than it should have been. Not only will you actually need thirty hours to complete your project, but on planet earth, each day also only has twenty-four hours in it. Thus, you need to start your project *before* the penultimate day of the semester, preferably.

In his book *The 80/20 Principle: The Secret to Achieving More with Less*, Richard Koch persuasively argues that since we are motivated to do what we enjoy, we should schedule those things into our calendars first and fill in the gaps with the lesser-desirable-but-still-must-get-done activities. Furthermore, Parkinson's law is the adage that "work expands so as to fill the time available for its completion", so if the difference is getting something done in one hour versus four hours (and if, being honest, this extra time will not make the quality four times greater), stick the task in a one-hour time block, get it done, and move on. Then, move onto the activities that really make your heart sing.

Now, I know you're probably thinking, *You just said three paragraphs ago to increase my time for study because I am underestimating what I need, but then you followed up with "you expand your work to fill up time"! Which is it?!* The answer is: both. Welcome to competing priorities, my friends! Yes, you do need to honestly assess how much time it will take to study for your classes and complete your long-term projects (without being *too* cushy with the time you're allocating for these tasks, or you will think you have time for procrastination). You will need to do this for each class on your semester schedule, *while* also remembering your work hours if you have those, your hours for sleep, mental downtime, club activities, religious activities (if you choose to partake in these), and physical training.

What does a rough budget look like? Let's take a look. We all have one hundred and sixty-eight hours in a week. If you are in twelve credit hours, that means you have twelve hours of class every week, plus a minimum of thirty-six hours of homework time. You also need to give yourself eight hours a night for sleep (which is fifty-six hours a week), eat three meals a day at an hour apiece (which is twenty-one hours a week), and engage in physical activity an hour a day for seven hours a week. Add in maybe ten for work, and we are up to one hundred and forty-two hours. In a seven-day period, that leaves twenty-six hours a week, or just under four hours per day, of "free" time.

Notice that in that list I did not include things like walking to and from class, doing your laundry, or buying groceries, nor were there any social activities, club events, seminars, or even the commute time to get to and from campus. *Your* calendar will need to include time for all those things, ample (but not infinite) study time, and time for real life (see Chapter 11 about taking a shower *daily*).

My next recommendation (stolen from my sister, who is a rocket scientist—yes, a real one) is to give yourself a heads up. One day, when I was snooping through her calendar (I'm the older, shorter sister, and therefore have the lifelong, God-given right to snoop), I noticed that a week before a major due date (like a paper or exam), my sister had written in big bold red letters, **Project Due 1 Week**, and the same for two, three, and four weeks out.

Brilliant.

No longer would I be ambushed when I turned the page in my calendar and, to my great horror, realized a project I should have been chipping away at for at least a month was now due in less than five days.

When Does Your Day Start?

"When does your day start?"

This is actually a trick question, and a favorite one of mine when I am giving workshops, because actually, your day starts the night before.

In consciously making decisions with the awareness that your day starts the night before, you really are setting yourself up for success. What this looks like will be up to you. If you have an important meeting or an interview first thing in the morning, I strongly encourage you to have your chosen outfit all laid out ready to go—each and every piece—pre-load your coffee pot (you should do this anyway!), and know ahead of time what you will have for breakfast. On days where it's more relaxed, I recommend at a minimum having your backpack organized and ready to go, along with your lunch or whatever snacks, a water, and a coffee thermos.

For reasons known only to those who study temporal mechanics, if you leave something to wait until the morning when you're rushing out of the house, it will take you twice as long to do it. If you do it the night before, it takes a fraction of the time and alleviates any stress. The math works itself out. Choose to plan ahead.

Study Appointments

Study appointments come in three different forms. The first is when you make an appointment with a couple of classmates from a specific class, like Calculus 1 or Physics 1, where you get together and work through problems and discuss them as you go. Do not be the Chatty Cathy leading people off-topic in these conversations, or you won't be invited again!

I recommend three to four people in a group like this. Any more than six and it gets unwieldy, and the propensity for distraction increases exponentially.

For this kind of study appointment, you will want to take over a table in the library (or one of the study rooms in the library) so that you are free to work and discuss without being a major source of distraction to those studying around you.

These appointments are a great way to get to know your colleagues and classmates, to help them with problems that you have figured out, and to receive help when you are struggling.

The next kind of study appointment is the one where you and a classmate just agree to meet at a library or a coffee shop or somewhere quiet because you both have work you need to do, and there's no talking. This kind of study appointment is important because when you're studying STEM subjects, you need to sit down and be able to process the information and solve problems on your own, as that is typically how you will be tested.

Both kinds of study groups are important, and serve their own purposes.

The reason why I always liked making a study appointment for a specified amount of time with a study partner is 1) it prevented me from procrastinating on a homework assignment, and 2) it also meant I only had an hour and a half for which I had to sit there and work, and then I was free to get up and go do something else. This was really helpful in getting me to just sit still and study.

The third kind of study appointment is a little different, and it's one that helped me with my writing assignments. I would look ahead in my calendar and determine when my writing assignment was due, and then back up my personal deadline to two days before that.[10] Once I completed my writing assignment by my personal deadline, I'd turn it in to the writing center to be reviewed, who would go through it with several red pens, bleeding all over my writing, correcting it improving it and helping me become a better writer. I would then take it back to my room and make the edits and adjustments that they recommended. This improved my writing and my grades significantly.

Personal Time

It is vital that you put personal time on your calendar and stick to it. I take one full day off every Sunday. I also take Friday afternoons off, starting at

---

[10] I would like to tell you I was more on the ball and could back it out a week, but that never actually happened.

about 2:30PM. This is a leftover habit from my time in sales, when we would put in ten- and twelve-hour days Monday through Thursday. By the time we arrived at Friday afternoon, we'd already have completed a full work week. Plus, no one we'd sell to wanted to see us after 2PM on a Friday, anyway.

I have continued with this in my academic and subsequent professional life. I also generally work on Saturdays, so taking Friday afternoon and evening off doesn't cause me any guilt. Some people will take a Wednesday afternoon or Saturday and Sunday in their entirety, which also works well. No matter how you decide to work this, I strongly recommend you book time off and do not fall into the mindset that working seven days a week is the best way to become successful (see *24/6* on this chapter's Recommended Reading list). We were not designed to work long hours, and certainly not seven days a week. You will notice a significant point of diminishing returns if you do so.

I will also tell you that there will be people around you who feel indignant that you have the audacity to take time off and refuse to schedule over it. They will not view your personal time as significant and important, especially if it interferes with what they are trying to do. My suggestion to you is, don't bother explaining yourself or justifying your time off. Just take it. Do not apologize for it. The sooner you put this into practice, the better. When you know you have time off that's dedicated to whatever you want it to be soon, whether it's going outside, working on a personal hobby, watching the next episode of *Star Trek*, or just staring at a beige wall because your brain is so fried, everything becomes so much easier to manage.

You absolutely must take consistent and regular time off for yourself. Remember, no company or a school program is contracted to look out for you, so *you* need to look out for you. Burnout and exhaustion are real, and they are particularly abundant in competitive college STEM programs.

The reason why I recommend having a specific day and time that you allot for your time off is because it will encourage you to get your work done in short order and put the books away. Plus, there is nothing like a deadline to finish your work, and most of what you do in college and in your professional life will be able to be cranked out in short order, anyway. You read about Parkinson's law earlier in this chapter, so you know a project can expand to fill the time available—if you cut your time off and demand of yourself the discipline to finish your project so you can go forward and enjoy time off guilt-free, you will be doing yourself a favor. Block it on your calendar so even if someone else wants to hold a meeting at 4PM on a Friday, you can say, in good faith, that you already have it booked. You don't have to tell them that it's you that the appointment is booked with!

Morning Routine

There are books upon books, articles upon articles, podcasts upon podcasts, and likely even LinkedIn blogs upon LinkedIn blogs, on the best possible morning routine. We'll discuss this in more depth in Chapter 11, but for now, know that in the end, the best possible morning routine for you will be unique to you. It just needs to work for you; that's the only criteria (well, sort of—more on that in a second). Here are some recommendations to consider when setting up your morning:

- Pick a time that you can wake up at consistently. Many articles will talk about the value of waking up at 5:30AM, 5AM, or some even as early as 4AM or 3:45AM. I tend to wake up just after 5AM, and that's enough of a head start for me.
- Pre-make coffee. I pre-set my coffee maker the night before so that it's ready to go when I wake up.
- Do your Bible study (or another reflective/spiritual/self-improvement task) in the morning. I also read a chapter of whatever book I happen to be focused on (or study my language, or write).
- If you have a full schedule, work out in the morning. If your schedule is more open, save your workouts as a reward for the afternoon.
- Have breakfast, but not too early. I personally have no problem with taking it with me out the door.
- Practice your personal hygiene and still get to campus or work without

risking life and limb and arriving stressed out and flustered, with feathers flying everywhere.

Above all, remember that just giving yourself even five extra minutes will make a world of difference.

Evening Routine

Curiously, I don't see much written on evening routines, even though these are just as important as morning routines. Again, yours will be unique to you and what you need (and will be covered in more depth later), but here are some things to consider for now:

- If you are a coffee drinker and have a programmable coffee maker, I recommend pre-setting it so it is one less thing for you to worry about in the morning.
- Have your hygiene practices accounted for (teeth brushed and flossed, face washed and moisturized, etc.).
- If you have a pet, you may need to let them out to relieve themselves.
- Pack your backpack and assemble all your homework to ensure everything is ready for the morning.
- Make sure your lunch is made for the next day.
- Take your vitamins if you haven't done so already.

I cannot study right up until bedtime and expect to have a good night's sleep, so when I'm building my schedule, I have to plan to shut down about an hour before I plan to go to bed.

One of the benefits of being in the science field is the fact that I have taken notes and made observations about what works best for me (the habits of a STEM student!), and I strongly encourage you to do this, too.

When you are assembling your calendar (including your morning and evening routine), it will be largely up to you to decide what you do and when. Then again, you must still operate within certain parameters set by society. You may want to get up at 10AM and go to bed at 2AM (I certainly have known students who want to live like that!), or perhaps, like Mitchell Feigenbaum (the father of chaos theory), you prefer a twenty-six-hour day to the standard twenty-four. However, most of society does not work that way, and if you want to make it through college, you need to adjust your schedule somewhat—assuming you still want to view it as practice for the professional world, that is (which I hope you do!). Keep this in mind when forming your morning and evening routine (and your daily schedule as a whole).

Creative Energy and Decision Making

We all only have so much creative energy to spend every day. Remember, this is what separates us from computers, and this is what will be your most valuable asset during and after college.

How you choose to spend your creative energy will determine whether you are depleted (the result of decision fatigue) or invigorated. You will also have to decide what decisions you want to invest your energy in and those you want to be rote so that you can spend your creative energy elsewhere.

A well-known and notable example of sartorial decision making is Steve Jobs. He wore a black turtleneck (or mock black turtleneck) and jeans all the time so that he did not have to decide, or spend any of his creativity, on what he was going to wear that day. That decision was already made for him. While I cannot (well, will not) speak to the current billionaires in Silicon Valley and ask if their T-shirts and hoodies are also a function of their wish to conserve creative energy (or if it's simply a lack of interest in how they dress), what I can say is that these everyday decisions crop up throughout the day, and for a variety of reasons. For example, I have no interest in designing and developing ideal workouts, complicated meal plans, or the like, and I know from personal experience that if I chew up all my creative decision making on little and (comparatively) insignificant to-do list tasks, I won't have the creative energy I need to go after and solve the much bigger and more interesting problems that arise.

As you are lining up your day and sifting through the decisions which make you productive versus the decisions which make you effective, take a good, hard look at how and where you are spending your creative energy.

If you really do not care, by all means, wear a black turtleneck and jeans every day. I personally need a little more variety than that, but not much. Your gifts to the world will be in solving these greater problems, like globally accessible energy, clean water availability, national security, or space travel and settlement.

Recommended Reading

- *The 80/20 Principle: The Secret to Achieving More with Less* by Richard Koch.
- *Cheaper by the Dozen* by Frank B. Gilbreth Jr. and Ernestine Gilbreth Carey.
- *Essentialism: The Disciplined Pursuit of Less* by Greg McKeown.
- *Getting Things Done: The Art of Stress-Free Productivity* by David Allen.
- *Write It Down, Make It Happen: Knowing What You Want and Getting It* by Henriette Klauser.

Strategies for Success: Organization

- Know the difference between productive and effective work. Arrange your day in such a way where you are effective.
- Keep your binder in order: no loose paper, each page dated and in its place.
- Buy the best quality tools you can afford and maintain them.
- Start your day the night before.
- Build consistent, effective morning and evening routines. Stick to them.

- Save your energy for creative problem solving. Turn rote decision making into routines to spend minimal creative energy.

7

Paying for School

Learning is not attained by chance, it must be sought for with ardor and attended to with diligence.
—ABIGAIL ADAMS

SOME OF YOU MAY HAVE had funds set aside by your parents or grandparents to pay for school, and won't have to worry about this section. Most of us, however, are not in that boat, and need to find a way to cover our expenses—which, mind, are more than just tuition, room, and board. You will be surprised at how expensive your books are every semester. Plus, there are now fees that universities do not advertise but can add to your semester bill (and believe you me, they are considerable sums). There are, of course, everyday life expenses that also need to be accounted for, like your vehicle, cellphone, or health insurance.

There is no getting around the fact that higher education is expensive, but if you are diligent, I believe it is still a worthy investment and recommend taking a very disciplined approach to your money while in school. This will serve you well over the course of your lifetime.

Money Management

I have certainly made mistakes in this regard, and I know for a fact I used student loan money to buy lattes more than twenty years ago that I am still paying for today. I write this chapter because I do not want you going down the same road.

There never seems to be enough money at the end of the semester. To give yourself a little safety net, you may consider building guardrails for your money management. For example, I know of one student who paid her bills up front as soon as her financial aid came in, meaning she paid her rent in four-month increments so she didn't have to worry about making that large payment every month. You may also choose to put your "food" money or "fun" money for the semester on a prepaid debit card. Once it's done, it's done, but you don't have to worry about overspending your budget. You may decide to rent a bit closer to campus and walk so you do not have to worry about paying for a campus parking permit.

Looking for ways to take care of your money is a solid habit to build in college and carry with you for the rest of your life. Besides, worrying about your finances will take away from the time and energy you would otherwise be spending on your studies or on socializing, and we won't accept that!

Safeguarding Your Financial Future

You have several options that will allow you to pay for school, which we will talk about next. Before we do, let's talk about one more critically important part of financing your education: you will need to pay the loans back.

Let me say that sentence again:

You will need to pay the loans back.

If you finance your education (and by that I mean sign a contract with a lending agency), you personally are on the hook to pay that loan back, just like you would a mortgage, a car loan, or a credit card.

I am not opposed to student loans and have some myself, but I made certain that my investment in my education went into degree programs that would provide a salary that would allow me to pay those very loans back at a later date. With this in mind (and very few institutions are willing to say this next statement, so I will), it is essential that you look at the salary ranges for careers that will be possible for your chosen degree. Do not take out $250,000 worth of loans for a career with a starting salary of $25,000. You will never get those loans paid off on that salary. Ensure you can afford the payments based on the starting salary of your chosen career field. Moreover, before you sign, make certain you know what the monthly payment will be, and draw up a rough budget that includes all your living expenses.

Now that we have covered that difficult (but real-life) issue, let's look at the four primary ways in which you can pay for school.

Student Loans

In the U.S., student loans come in two general flavors: federally backed loans and private loans.

Federally backed loans typically have lower interest rates, longer payback periods, the option for income-driven payment plans, and the possibility of forgiveness in certain circumstances. They are the most common way of paying for higher education. You do not need to start paying these loans until after you graduate or have been out of school for a certain amount of time. Subsidized loans do not accrue interest while you are in your program, but will start as soon as you graduate. If you go back to school for advanced degrees, these loans can generally be put into deferment, and the interest will accrue until you graduate with your next degree, but you won't need to make payments while in graduate school. If you can keep up with payments, I recommend doing so.

Private student loans are much more like a loan you would take out to buy a car, for example. The interest rates are less favorable, the payback terms are shorter, and they are ineligible for any of the perks that come with federal loans. Also, interest continually accrues with these loans. There is no deferment, either. Use these loans as a last resort... or, better yet, not at all.

Scholarships and Grants

Scholarships are a wonderful way to help pay for school, and I cannot overemphasize how important it is to seek out scholarships and grants related to your program of study, university, clubs, association memberships, and hobbies. There are also occasionally alumni scholarships that are specific to your hometown.[11] Scholarships are competitive, however, and with that in mind, I have three recommendations for you on your hunt.

First, operate on a volume basis. Do not just apply for one, even if you are the ideal candidate and assume you will get it. A good rule of thumb is that ten percent of the scholarships you apply for, you will get. Accordingly, I recommend applying for at least ten different scholarships every year. If your heart just sank at that impending amount of work, don't give up just yet!

This brings me to my second point, and that is: at the beginning of every scholarship season (which is usually in early to late spring), spend a good deal of time working on one essay. You can scan through and see the general essay requirements for several scholarships, and that will give you a really good idea of what to spend your words on. I say this because I spent several hours polishing, refining, and editing my essay and having it reviewed, and walked away with several thousands of dollars in scholarships. That year was a particularly good year, and I spent nearly eight hours on that essay to make it perfect. This was absolutely worth it when you look at the dollars per hour that I earned with it!

The last piece of advice I have is, follow the instructions to the letter.

---

[11] What I have never found is the mythical scholarship for left-handed people. Over thirty-two semesters, I have searched high and low and never found it—I'm left-handed, you see—so if you do find that left-handed scholarship, please send it my way so that I can see it!

Because scholarships are so competitive, committees look for (what feel like arbitrary) reasons to immediately eliminate many applicants, so they automatically cut those who did not follow the instructions. Some will have weird or really specific instructions like margin size and font size and type— a list of this or that—so whatever they are asking for, just do it.

If you are unsure about whether you would qualify for a scholarship because the instructions are unclear, take the time to call and find out and explain your personal circumstances. This can save you time and money. If you have to submit a paper application and pay for it to be certified, you will want to be certain you have a reasonable chance of winning it (that one in ten chance we spoke about earlier). This happened to me with a scholarship that I was ready to apply for during graduate school: when I called the committee to confirm my eligibility, it turned out I was not eligible. They wanted students who were drilling water wells, not petroleum wells.

If it is not the right fit for you, shrug it off and move on.

Summer Internships and Co-Ops

Summer internships and co-ops, or a paid semester internship (where the work you do is also for a grade), are excellent ways to achieve several ends with one mean. First, you are paid—and usually pretty well!—when you work on an internship or co-op. Second, they give you the chance to interview the company (and thus also the industry and the type of work you might be involved in professionally). Third, internships and co-ops give you a foot in the door with a potential employer after graduation. Fourth, they generally have you work on a project over the course of the summer, and these projects can and should be included on your résumé; this work experience will demonstrate your increased knowledge and expertise. Lastly, you will meet professionals on many different rungs of the corporate ladder, and these professionals can serve as career guides, mentors, references, and advocates.

If you are a freshman looking for work for your first summer, do not be discouraged if the responses are lukewarm. Get dressed up in your best suit, print off your business cards, polish your résumé, go shake hands, and sign up for interviews (we'll cover all of this in more detail in Chapters 12, 13, and 14). Depending on the economy, employers will either want to hire just about anyone with a pulse or will be incredibly competitive, with most positions automatically going to incoming juniors and seniors. Regardless of the economy, get out there and work the career fair. None of your efforts will be wasted. This is a prime opportunity to practice your hard-won soft skills (see Chapter 9), and you just might land yourself on the radar of someone who is hiring.

Work Study

In the U.S., students have the option of participating in what is called "work study". This program allows students to work on campus, ideally in a position related to their field. Work study is part of the financial aid package given prior to the start of term. I recommend visiting your financial aid officer about this option and keeping an eye out for positions on campus that would both pay well and relate to what you are studying. An added bonus to this approach is that you build your network on campus and have someone who can write letters of recommendation for you. Depending on where you do your work study, it may even open doors for future jobs or careers in your field.

Recommended Reading

- *The Richest Man in Babylon* by George Clason.
- *Rich Dad, Poor Dad* by Robert T. Kiyosaki.
- *The Big Sister's Guide to the World of Work: The Inside Rules Every Working Girl Must Know* by Marcelle DiFalco and Jocelyn Greenky Herz.

Strategies for Success: Paying for School

- Spend time writing and polishing your annual scholarship essay. Customize as necessary.
- Follow all scholarship and grant application directions to the letter.
- Remember that internships and co-ops are two of the best ways to pay for school and prepare yourself for the professional world.
- Know what you are signing and for how much if you take out loans for school.
- Even if you can only work five to ten hours a week, it is worth doing for the experience, résumé-building, networking potential, and money to cover minor expenses.

8

Reading

If you haven't read hundreds of books, you are functionally illiterate, and you will be incompetent, because your personal experiences alone aren't broad enough to sustain you.
—GENERAL JAMES MATTIS

I N MY FRESHMAN YEAR OF high school, I joined the debate team as a Lincoln-Douglas (L-D) debater.[12] At the first meeting, our coach asked a few of the older, more experienced teammates to give advice that was sure to help us become champion debaters. I remember nothing from that experience except one piece from a tenth grader. Heather "The Hatchet" Hackett lectured us on how we should always be reading. Specifically, "Read a book related to your debate topic," was her advice. It did not have to be about the precise issue being debated, but it did have to be linked in some way.

Lincoln-Douglas debaters argued a new topic every month, and policy debaters argued one policy prompt for the academic year. Her reasoning for

---

[12] Disclaimer: I did so because I came home and announced to my mom that I was trying out for cheerleading. She replied, pointedly, "You can be on the debate team." Thus ended my cheer career.

this advice was that not only will books provide additional ideas to argue with (or against), but you will have a better understanding of the topic than your opponent. Plus, it impresses the judges (which it did).

I kept up with her advice throughout high school, and after graduation, I started applying her rule to traveling (or moving, or a new degree program, or a new job). No longer did I need to squash opponents or impress judges; I just liked knowing more about where I was going and what I was doing.

Reading (or not reading) is generally correlated with your educational background, but (as we saw with Benjamin Franklin) not absolutely so. A strong reading habit is, however, most certainly correlated with your *success* in life. Warren Buffett spends upward of five hours a day reading, and his reading includes books, newspapers, and journals.

Books specifically are an interesting category, because the advent of eBooks and audiobooks has broadened "books" as a concept and led us to question what "actually counts" as reading. For our purposes, any of these formats works. I read actual paper books when I am at home in the morning and in the evening, I listen to audiobooks when I am working out or cleaning the house, and I will read eBooks on occasion on my phone when I am traveling. The most important thing is that you are reading, and from a wide variety of sources.

During my sales career, I called on a physician's assistant who was also a new mom. Her practice kept her days full, and her family kept her personal time busy. She mentioned to me during one of my calls that she spent fifteen minutes every night before bed reading a technical paper. She wanted her medical practice to be at the forefront of health and science. I have no doubt her patients continue to benefit from her personal commitment to daily learning. Make reading a part of your daily habits.

Do not stop reading when you graduate from college. Do a quick search on people who have risen to the top of their professions, and you will see more often than not that somewhere buried in their daily habits is a commitment to reading—or, in other words, a commitment to learning. So, make reading part of your regular habits, even if it is just for fifteen minutes before bed every night, or a chapter every morning. Over time, the accumulated effect of continual learning and the stretching of your mind will have a profound impact on how you view and understand the world and your contributions to it.

One last note on reading: books and reading helped win World War II. Our minds are shaped by what we put into them, and the U.S. military knew that before we could win the war on the battlefront, we needed to win the war of ideas—liberty, democracy, and freedom, to name a few.

Books

What does this have to do with college? And how can you possibly be expected to read more when you can barely finish all your homework? No matter where you are at in school or where you are going after graduation, you can and should pick up a book. Wikipedia pages and YouTube videos are helpful if you want to get a cursory understanding while still digging in a little bit, but to find a richer explanation, deeper understanding, or funny insight into your study topic, read something fiction—even science fiction!

Recently, I sat around the lunch table at a highly technical conference with fellows I had never met. Did we dissect the nuances of expert presentations? No. We argued over the storylines from *Dune, Stranger in a Strange Land*, and *Brave New World*, as well as all things Tolkien.

People who read are like people who belong to a secret club: we seek each other out and talk excitedly about our latest read. We can see patterns through history, and we marvel at the dogged persistence of heroes and heroines (real life and fiction). We ascend faster and more often to the tops of our fields.

At the end of this chapter, I have included several books related to STEM fields. They are all excellent. I did not include any science fiction, however; I will leave that recommended reading list to the robust collection I'm sure you already have.

In Appendix B at the end of this book is a starter personal library list that provides a variety of subjects (mostly nonfiction) for you to read about. Each one is related to a section in this book. You should be able to find all of these in paperback and used form at your local bookstore. Buy one at a time and read a chapter a day if you can—or a chapter a week, if that works better. Take one or two pieces that you learned in your book and apply it directly to your life. Yes, there is great satisfaction in seeing your bookmark march across the pages as you close out a chapter every day for your reading, but reading books is meant to have an impact on who you are.

Industry News and Trade Journals

For those of you who are planning on going into the professional world after graduation, I recommend that you (in addition to books) start reading industry news and trade journals. As you start out, these will all likely be in digital format, and this is nice, because you won't have piles of back issues that you need to figure out where to recycle building up in the corner of your dorm room. To get started, I recommend going through your university library to check out online subscriptions.

Whether you are engaging with these texts in digital or paper form, the benefits of reading industry news and trade journals lie in you garnering an increased understanding of the field in which you want to spend a good portion of your professional life.

Initially, you may not understand much of what is going on—the names, problems, and concerns will seem odd and hard to relate to—but if you continue scanning the topics and headlines and reading an article or two in every issue as you go through your college career, you will be stunned at how much background information you pick up about your industry. You will also understand who the big players in your industry (and the up-and-coming smaller companies that might provide exciting employment opportunities) are, and you will come to learn of the issues relating to state or federal legislation. There are literally hundreds of trade journals for every conceivable industry, ranging from automotive, transportation, and travel, to energy and technology.

If you are not exactly sure what you want to do after graduation, this is a good way to start perusing some of the different options before you. If you are in a STEM field, you are a natural problem solver, so keep an eye out for the kinds of problems that you want to use your hard-won brains to solve.

Knowing your industry news will also help when you are interviewing in the Fall/Spring Career Fairs, and it will make your education that much more relevant when you are sitting in class. At some point, you will be in an upper-level course, and suddenly, a problem you read about in your industry journal will jump out at you. This is when your studies get really exciting: you will be working on problems that actually have real implications for what is going on in your industry right now and with which you are already familiar.

For those of you who are interested in research and are focused more on pure science and mathematics, please see the next section on periodicals and peer reviewed journals.

Periodicals and Peer Reviewed Journals

This section should be treated with a light touch. Periodicals and peer reviewed journals are the inverse of an undergraduate education: where undergraduates study a mile wide and an inch deep, periodicals and peer reviewed journal articles are an inch wide and a mile deep—sometimes deeper.

As an undergraduate, you are still in the process of developing your discipline-specific language skills, so understanding this language will be one of your first challenges when you are reading articles of this sort. Peer reviewed articles are written with postgraduate language for good reason: this is how advanced researchers communicate with one another (although some could argue writing of this sort is done with deliberate obfuscation).

The next challenge lies in the fact that every discipline is studied and researched to a finer and finer tune, meaning one article does not cover a substantial portion of the overall discipline, but a fraction of a fraction of it.

As an undergrad, you are not expected to know the breadth and depth of your discipline to this degree. If you did, you would be a graduate student or a postdoc instead. I just wanted to introduce you to this section of literature because a portion of you will be going on to do graduate level and postdoctoral research, so knowing these journal articles exist and carry great importance for advancing the frontiers of scientific research and human knowledge is a good thing. Plus, exploring these journals can be great fun and interesting. You may read something that surprises you and changes your planned trajectory.

You should also proceed with caution for the opposite reason: as an undergraduate, it is easy to read one article and fixate on it for all the research you conduct, whether for an undergraduate research project or perhaps in an upcoming master's thesis. In these cases, one article will not suffice. Every thesis and dissertation requires a literature review, which means you will need to sufficiently survey *all* the literature relating to your study question (or at least a decent chunk of it).

Undergraduate professors do not expect their students to read periodicals and peer reviewed journals, nor for them to have the time to do this. If you want to go in and flip through some of the journals and see what people are researching and what may be of interest in the future to you, by all means, I recommend that you do so. Do not neglect your undergraduate studies and do continue reading the required and recommended portions your professors assign, however. Only if you find yourself wandering the stacks and wondering where the cutting-edge research might be going in your field should you go peruse peer reviewed journals.

Put It into Practice

Did you read as a child? Were you read to when you were small, perhaps before bedtime? Did you get lost in an adventure novel on a Saturday when you were supposed to be tidying your room?

Reading is one of the last places we can still go to for a much-needed break for very little money. So, why do so many people stop reading after childhood? It is hard to say, but a likely reason is that our technology hijacked our attention. Yet if ever there was a panacea to solve our problems, it is the book. We all need quiet time to be with our own thoughts. We need adventure. We need advice. We need laughter and grief, and even a little heart-thumping fear! We need to know what has previously happened on the other side of the world and in our own neighborhoods. We need to know how certain things were done so we can do them better next time. Like the good general said, if we haven't read hundreds of books, we are functionally illiterate.

But what if reading feels painfully challenging? First, make sure you're not just out of practice. The first time back in the gym after I've taken time off is definitely painful, but that's not because I'm doomed to never do a killer session again. Don't force yourself to read a heavy tome if it's been a while with a book; pick up one you remember that you liked and just enjoy it.

If, however, you have always struggled to read, please see your Student Success Center (every campus has one; yours may just have a different name) to get tested for a reading disability. If it turns out you have one, does this mean you are out of the secret club? Absolutely not. One of my favorite professors in divinity school had a terrible form of dyslexia.

Wait... a professor of theology had dyslexia?! Yup.

In Divinity School, we read a staggering number of pages daily, and this man made it all the way through his program and graduated—meaning, he also wrote and successfully defended his dissertation. He did this by having someone read the material out loud into a tape recorder, and he followed the words along the pages. *For years.* He put in an incredible amount of effort to accomplish what came far more easily to many of us.

He continued reading when he was well into his career, because he had built a habit that worked.

Now, with advances in technology, this process can be much quicker. If reading a book in a traditional format is difficult for you, go with audiobooks (either fully, or as an accompaniment to your physical reading).

As with all good habits, there is every reason in the world—some good, most bad—to not get after this goal. We are too tired; too busy; our eyes are tired and dried out from too much screen time; our brains are too frantic or too dull; we have more work to do; and on and on. Make the time anyway. No one ever said, "Well, we would hire this engineer [or scientist, or mathematician, or technologist), but they read too much." On the other hand, when a prospective employee has admitted to not ever having opened a book beyond what was required in school, employers have said, "Someone so opposed to learning is generally unteachable. Do not hire." I don't know about you, but I know what side of the fence I'd rather be on.

Recommended Reading

- *When Books Went To War: The Stories That Helped Us Win WWII* by Molly Guptill Manning.
- *Zero: The Biography of a Dangerous Idea* by Charles Seife.
- *Humble Pi: A Comedy of Maths Errors* by Matt Parker.
- *The Immortal Life of Henrietta Lacks* by Rebecca Skloot.
- *Longitude: The True Story of a Lone Genius Who Solved the Greatest Scientific Problem of His Time* by Dava Sobel.
- *Sea Change: A Message of the Oceans* by Sylvia Earle.
- *The Soul of an Octopus: A Surprising Exploration Into the Wonder of Consciousness* by Sy Montgomery.
- *The Coming Plague: Newly Emerging Diseases in a World Out of Balance* by Laurie Garrett.
- *The Emperor of All Maladies: A Biography of Cancer* by Siddhartha Mukherjee.
- *Guns, Germs and Steel: The Fates of Human Societies* by Jared Diamond.
- *Astrophysics for People in a Hurry: Essays on the Universe and Our Place Within It* by Neil Degrasse Tyson.
- *The Disappearing Spoon: And Other True Tales of Madness, Love, and the History of the World from the Periodic Table of the Elements* by Sam

Kean.

- *Hidden Figures: The American Dream and the Untold Story of the Black Women Mathematicians Who Helped Win the Space Race* by Margot Lee Shetterly.
- *The Boy Who Harnessed the Wind: Creating Currents of Electricity and Hope* by William Kamkwamba and Bryan Mealer.
- *The Invention of Nature: Alexander von Humboldt's New World* by Andrea Wulf.
- *Broad Band: The Untold Story of the Women Who Made the Internet* by Claire L. Evans.

Strategies for Success: Reading

- Read a book a month while in college. You don't have to go quickly; just make progress. Make this a habit for life.
- Choose a book that benefits your growth intellectually or professionally. Consider these books your mentors.
- Spend a little time every week in the library reading an article or two, a trade journal, or a peer reviewed journal from your industry.
- Reference what you've read recently in conversation with industry professionals or potential employers.
- Find other readers in your friendship circle and meet to discuss what you read.
- If you read sci-fi, do not waste the study opportunity: figure out what would actually work in the book (in terms of the science or engineering) and what would not. Start applying what you are learning in your courses.

9

Develop Your Soft Skills

People don't care how much you know until they know how much you care.
—THEODORE ROOSEVELT

W E WOULD ALL LOVE TO believe that it is the genius of our ideas—our technical acumen for hard skills—that will win the day. Whoever has the best idea will rise to the top! Soft skills are weak, right?

Actually, no, they are not. In fact, companies spend a considerable amount of money hiring the best possible employees and keeping them trained in line with the newest research on leadership, team building, and human organization (hint: these are soft skills). If all it took was buying computers to solve problems, that is exactly what companies would do. Computers cost a lot less than people, after all.

Companies want employees who are technically proficient *and* who can work well with others. Our creativity and our soft skills differentiate us from computers, robots, and AI (well, for now), and these human attributes are worth honing and developing in college.[13] You will absolutely need these

---

[13] See the end of this chapter for a list of soft skills you can include in your résumé or work on individually as you go through school.

skills in order to be successful in college and your professional life. We exist in a marketplace of favors, and that means interacting with our fellow humans. Do not neglect your study of soft skills!

There are many ways to develop soft skills, some better than others. The good news is, college is a wonderful place for you to learn and practice, partly because it is a low-stakes environment in which you can make mistakes. I, for one, strongly recommend that you make good, honest mistakes frequently.[14]

Any position that requires you to interact with people will give you the chance to practice your soft skills, whether tutoring, holding office in a student chapter or club, working in an office (with people!), serving as resident assistant in your dorm, or playing on a sports team. These will all give you ample opportunity to practice, refine, improve, and try again.

Working with people can be a challenge. We are not uniform and solvable like many of our equations and homework problems, and working with volunteers is a unique challenge. Being a leader on a team means additional responsibilities, and all these skills can be transferred to a postgrad career, the military, or graduate school. The ability to relate to others and communicate your ideas will be critical to your success in college and the professional world, and your ability to communicate complex scientific or mathematical problems in particular will set you apart in any crowd. Carl Sagan, for example, was a terrific science communicator who could bring the general population along with him in scientific understanding.

If you are still not convinced about the importance of soft skills, look at the difference between Nikola Tesla and Thomas Edison. Tesla, by any standard, was a genius. Edison... maybe, maybe not. Yet while Tesla died a pauper, Edison did not. The explanation is simple: one lacked soft skills; the other did not.

Before we dive in, it is worth remembering that honing your soft skills is meant to be fun. See the list at the very end of this chapter for skills you

---

[14] See the "Clubs and Associations" section in Chapter 12.

already possess and those which could use a little polishing. Go through the list and circle those you already have. Include these skills in your cover letter, job applications, or your résumé. Then, star the soft skills you would like to work on.

While all the skills listed at the end of this chapter are important, two skills that consistently rise to the top in my conversations with students are "leadership" and "public speaking", so we will spend the bulk of our time in this chapter on these two soft skills. We will touch on a few others that will help you navigate college life and beyond too, though.

Let's jump in and talk about the soft skills you can develop while simultaneously learning your technical skills and building up your résumé.

Leadership

It was not until I was asked to write a letter addressing my own leadership abilities that I really considered how I understood leadership. This soft skill is interesting, because there are so many different leadership styles; there is no one right way to be a leader (although there are certainly a few *wrong* ways to go about leadership).

To answer the question, "How do I understand leadership?" I could have logged onto my Amazon account, ordered half a dozen books written by leadership professionals, constructed a thesis statement, and defended my deftly articulated argument in one to two pages.

Or I could write about my lived experience.

Leadership is our influence on those around us. Do we inspire those next to us to stand up and make the hard decision, or to take the easy path? Do we uplift and promote those around us, or do we create a toxic, backbiting environment?

When it comes to leadership comportment (behavior), we have three models from which to choose: scarcity, mediocrity, or generosity. Transient leadership (our temporary effect on our colleagues) may stem from any one of these three models, but only one, I believe, effects positive and lasting change. See if you can guess which one.

- The scarcity model encompasses the idea that a finite amount of success exists (meaning if you do not get it, someone else will). This leads to a corrosive environment of backstabbing, pettiness, and even professional sabotage. Morale deteriorates, as does enthusiasm, energy, and *esprit de corps*. Scholarships, journal publications, and grant money may externally abound, but this model exacts too high a cost by decaying the interior life of the individual.

- The mediocrity model, despite sounding benign, insidiously degrades the quality of achievement. Sitting in one's office, the mediocre attitude mumbles, "I will do enough, but no more." Mediocrity refrains from inviting fellow students to step up to do a little bit better. Each student generation does just a little bit less than the prior generation, curbing the pursuit of excellence and, eventually, excellence itself.

- The generosity model requires the action of at least two people: a giver and a receiver. The idea here is that we cannot advance in isolation, but must interact with one another to make progress. Through interaction, generosity assembles, builds up, and strengthens. When we offer a compliment, a word of encouragement, or a "learn from my mistakes" comment, we build solidarity with our fellow students. In other words, we freely give the best of ourselves.

I have been on the receiving end of the generosity model so often that it will take me more than a full lifetime for me to break even. My mindset is, I can afford to be generous because I know that there are opportunities for all of us: women and men; scientists and engineers (and mathematicians and technologists); undergraduates and graduates; new career, second career, end of career... There are so many opportunities for success and personal accomplishment that I have no reason to be miserly with my support, and I suspect the same goes for you! You did not land in this program all by yourself: someone helped you. Some people helped you with your homework, wrote letters of recommendation, or put your name forward for scholarships or awards. So, take a look around and see who could use a little nudge of encouragement. Why not cheer them on, give them the best of what you have, and pump your fist in the air at their achievements?

It has been a great privilege for me to write letters of recommendation for graduate students for research projects, graduate school, and promotion over the course of the last three short years; to encourage and vote for shy leaders in student organizations; to critique technical presentations; to regale students with all my mistakes so they can take off faster, better, and smarter than I ever did. You will never regret lending a helping hand to someone striving to reach the position you're in.

Public Speaking

Like writing, the ability to communicate through oral communication is vital to your success as an engineer, scientist, technologist, or mathematician. All the glorious ideas in the world will have no impact if they are not sufficiently communicated to the wider public in a way that makes sense to laypeople.

There are a few traps we fall into when presenting our information, and I certainly am the chief of sinners when it comes to these mistakes. The main one is, we are so excited about our work that we assume everyone in the audience wants to know every excruciating detail along with us. However, as one of my advisors reminded me (on more than one occasion), everyone is excited to see the baby, but nobody wants to hear about the labor pains. If it took you ten thousand tries to get through your research, and on the ten thousand and first try you finally had a breakthrough, I'm sorry to tell you that nobody wants to hear about the agony that was the ten thousand tries that preceded your success. I know you want to have the entire audience share in the drama of your angst and frustration, but they simply do not want to hear it. Tell your roommate, tell your lab partner, tell your best friend, or even send me an email and share your story with me, but when standing up to present the dramatic results of your experiment, keep it focused on the success and the finished product.

Presenting from PowerPoint Slides

We like to put as much information as possible on PowerPoint slides, and this is what contributes to "death by PowerPoint syndrome". My strongest recommendation here is, only include figures on your PowerPoint slides (besides the initial title page, agenda, and final page, which will note the conclusion or ending questions). You've done the research and you know your material, so you can speak into the figures on the slides without scripting everything out for yourself onscreen.

There's also the challenge of font size for axes on graphs and charts. Sometimes, there simply is no getting around small font. One way to help your audience in this respect (and I personally have had tremendous success with this technique) is to pre-describe the graph before you get into its contents. First, tell your audience what is on the Y-axis and the X-axis, including the units (we read from left to right, so it will make sense to your audience for you to start with the Y-axis rather than the X-axis). If there are colored bars or trend lines, let them know what those are, and then go into the details of the graph or chart and why you have it in your presentation.

Take the time to paint the story about each slide so that the relevance of each one is clear to your audience. While the story may seem obvious to you, this is because you have been studying the material and working with it for quite some time. It is likely that your audience is seeing this information for the very first time and has not had the chance to absorb and process the materials like you have. So, help your audience out and explain why you are telling them what you are, so that when you finish, your argument has been driven home and your presentation has been a successful one.

Communication Apprehension

There are physiological challenges that arise when we are presenting, and chief among them is communication apprehension. Everyone has communication apprehension; it's just a matter of developing the skills needed to set it aside so you can present without your voice shaking.

I notice that when my chest tightens up, of course my voice starts to shake, and then I sound nervous, which makes me *actually* nervous, which causes my voice to shake even more. To prevent your chest from tightening up in the first place (thus triggering this entire cycle), swing your arms forward and backward as far as they can go, to get your body to loosen up. Take deep breaths. Remember that this physical reaction is perfectly normal, because standing up in front of people truly is a vulnerable position to put yourself in.

Over time, you will experience communication apprehension less and less. There will always be the initial surge of adrenaline that catches you when you first start a speech, but know that this is normal and that you are in good company.

I also have trouble with my speech rate. I tend to talk quickly anyway, but when I am amped up on caffeine and adrenaline (and am excited about my presentation material), I can speak like an audiobook on double speed. This is exhausting for the audience to listen to. I'm still working on slowing down!

Some of these habits are lifelong practices, so do not worry if you cannot get it exactly right straight away. No matter what happens when you stand in front of your audience, remember you are the expert. This should bring you a great sense of relief and peace. The audience is there to listen to you.

When you are reviewing your speech or presentation, the temptation is to review it silently in your head and just scan over the words. This is not the same as practicing your speech. This leads us to...

Practice Out Loud

Practicing your speech involves (or should involve!) you standing up and speaking the words out loud. I once had a manager who used to describe this as "getting the marbles out of your mouth". I would sit in my car "practicing" a sales pitch silently in my head, and he was insistent on me speaking it out loud. Initially, I thought this was a frustrating waste of time because I was legitimately reviewing the material in my head (or so I thought)... and then I went into my sales call and choked.

It turns out that reviewing in your head is not the same as speaking out loud. When you make yourself actually talk through your speech and "get the marbles out of your mouth" (even just two or three times before you stand up for the actual presentation), you will be stunned to find how much better you do when you are in the front of the room and all eyes are on you.

Practice smiling while you speak from time to time. It is easy to be concentrating so hard on your subject matter that you start furrowing your brow and tensing your jaw, and this has the unfortunate effect of making you look cross in front of your audience. You want them to know that you are excited about your work that and they should be excited with you, so show off those pearly whites!

There are many good books and videos on public speaking. I personally would avail myself of the books in the Recommended Reading list at the end of this chapter. Then, as with every other habit in this book, practice makes perfect. Consider joining ToastMasters to give yourself a live audience, and just keep practicing!

Soft Skills in Demand

- Adaptability. The business world changes *a lot*. Companies are bought and sold, leadership changes over time, and new management ideas take hold. The first principles you studied in your STEM courses will hold fast, but not much else in this fast-paced world will. Get your work done despite any external chaos, like water off a duck's back.
- Business etiquette. Know how to properly address those who outrank you, introduce a colleague to a new acquaintance, and make conversation at a business dinner (and which fork to use). There are subtle graces we can employ and deploy to make those around us feel at

ease. When clients and business partners enjoy being around you, you will be the first your boss sends out to represent the company. Well done!

- Being collaborative. Everything is done in teams in the workplace. Employers are not looking for hotshots, showoffs, or any other form of arrogance. Be the kind of person that you would want to work with: open minded, willing to contribute, knowledgeable about current subject matter, honest in your limits, friendly, and generous.

- Dependability. Be the person who finishes their work, and does so with aplomb! While we all occasionally fall behind, those who stand out are those who do not need to be asked repeatedly for their work; who submit it for review first and then on time, with few (or no!) mistakes. If you are coming up to a deadline and know you won't make it, immediately reach out to your team or your client and own it. Give them a best estimate for submission, without long explanations. No one cares why it will be late; they just want to know when to expect the work.

- Empathy. No one wants a cold, heartless boss. Movies and books are full of that character type, and they always get what they deserve in the end (and for that, we cheer!). We do not always know what is going on behind the scenes with someone (sickness, death in the family, financial duress), nor is it our business. Kindness and empathy should take centerstage in our daily interactions, as we may well be in that situation one day. Show grace. If you have the bandwidth, ask if you can help. Don't thump on your colleagues. It's been my experience that people are honestly doing the very best they can.

- Being ethical. How many CEOs and senior members of leadership have we seen fall because of their unethical behavior (either toward subordinates or finances)? These people are often fired for their lack of character. Those who are lower on the corporate ladder suffer the same fate, mind; they just do not make the headlines. How we represent our time on our timecards, to how we handle interoffice relationships, to how we manage corporate and client money, and everything in between, bumps up against ethics. Maintain a high standard, and you won't have any regrets. If your ethical standard is bumping up against company expectations, find a new company.

- Punctuality/timeliness. This soft skill can readily distinguish you from most. People rarely plan ahead enough to arrive on time at work, meetings, and the like, never mind early, yet they know how to arrive at the airport early enough that even with a long line through security, they won't miss their flight. This same awareness is rarely employed for more "normal" day to day deadlines (which are way less difficult to figure out). While the odd unexpected delay can happen *on occasion*, in general, we are well served when we apply the same level of timeliness to clients, colleagues, and supervisors. If you are routinely late, it may not be mentioned, but it will always be noticed. Punctuality (or tardiness) is a choice.

- Problem definition/solving. The biggest surprise for me when I entered the workforce was the distinct lack of problem definition (and, by extension, problem solving) skills in clients. More often than not, clients will approach me or my colleagues with a problem with a vaguely defined scope, for which there will be little hope of solution. If you possess the talent for seeing and clearly defining the problem and then proposing the solution, you will win the day. Keep this skill sharp and at the ready, and you will never lack for good work.

- Independence. One of the most exasperating tasks for a supervisor is having to constantly handhold employees or subordinates. Asking questions for clarification is one thing; constantly asking what the next step needs to be, or, worse yet, having to be constantly reminded by your supervisor to take the next step, is unacceptable. Being able to prioritize your work, manage your time, and produce consistently solid work is all part of being self-directed. This does not mean that if you are unclear about the project or need better direction for the next step that you shouldn't just ask the question, however; part of being sufficiently self-directed includes knowing when you are in territory that falls beyond your abilities.

- Strong work ethic. The good news for you is that the fact that you are making your way through a STEM degree or program (while remaining engaged in extracurricular activities) means you already have a fine work ethic. Show up to work on time, ready to roll. Do good work throughout the day. In the evenings or on weekends, if it is urgent, then continue your tasks, but think long and hard before going much over fifty hours a week of total worktime.[15] Every virtue carries within it the seeds of its own destruction. A strong work ethic can quickly turn into workaholism and burnout. Keep on your guard about that. Then again, you already know that someone who keeps trying to shorten their workday or work week will not last very long. Put a good day's effort in for a good day's wage.

Recommended Reading

- *Dare to Lead: Brave Work. Tough Conversations. Whole Hearts.* by Brené Brown.
- *Pitch Anything: An Innovative Method for Presenting, Persuading, and Winning the Deal* by Orin Kleff.
- *The Well-Spoken Woman: Your Guide to Looking and Sounding Your Best* by Christine K. Jahnke.

Strategies for Success: Soft Skills

- There are as many leadership styles as there are books on leadership. Whether you lead from the front or back, it does not matter: how you treat people matters.
- You will give presentations throughout college and well into your career, so start practicing now.

---

[15] Forty hours a week is expected for a full-time position, but I understand that early in one's career, it is exciting to put extra effort in and develop one's craft.

- Take the time to polish your slides. This will give you confidence.
- Practice your presentation out loud. This will give you even more confidence.
- Nobody wants to hear about the labor pains; they only want to see the baby.

Here are ninety-nine soft skills for your résumé. Soft skills include communication, social skills, and integrity. Whether you need a noun, a verb, an adverb, or an adjective, use the following soft skills to tell employers what you bring to the team.

This list is inspired by the other "soft skills" lists I came across online while writing my blog on this topic. It is not complete by any means, but should give you a solid start!

1. Accepts Feedback
2. Accountable
3. Adaptable
4. Adjustable
5. Agreeable
6. Articulate Speaker
7. Aspirational.
8. Assiduous
9. Attentive
10. Business Etiquette
11. Calm
12. Collaborative
13. Committed
14. Common Sense
15. Communicative
16. Competent
17. Confident
18. Conflict Management
19. Conflict Resolution
20. Considerate
21. Cooperative
22. Courteous
23. Creative
24. Credible
25. Critical
26. Decent
27. Dependable
28. Determined
29. Diligent
30. Driven
31. Effective
32. Efficient
33. Emotionally Intelligent
34. Empathetic
35. Encouraging
36. Enthusiastic

37. Ethical
38. Facilitative
39. Flexible
40. Focused
41. Follow-Through
42. Fortitudinous
43. Friendly
44. Gives Feedback
45. Happy
46. Helpful
47. Humorous
48. Industrious
49. Influential
50. Initiative
51. Integrity
52. Interpersonal Skills
53. Intrepid
54. Kind
55. Know When to Communicate
56. Leader
57. Lifelong Learning
58. Likable
59. Listening Skills
60. Mature
61. Mentor
62. Nurturing
63. Objective
64. Open-Mindedness
65. Optimistic
66. Organized
67. Patient
68. People Skills
69. Perseverance
70. Personable
71. Persuasive
72. Polite
73. Positive Attitude
74. Presentation Skills
75. Principled
76. Problem-Solver
77. Productive
78. Professional
79. Punctual
80. Research
81. Resourceful
82. Respectable
83. Respectful
84. Self-Controlled
85. Self-Directed
86. Self-Disciplined
87. Self-Motivated
88. Sociable
89. Social Skills
90. Solution-Focused
91. Supportive
92. Teachable
93. Team Player
94. Time Management
95. Versatile
96. Virtuous
97. Willing to Change
98. Willing to Learn
99. Work Ethic

10

Social Person, Social Life, Social Media

You shouldn't take life too seriously. You'll never get out alive.
—VAN WILDER

THIS WILL BE THE SMALLEST chapter in this book, but its contents may have the biggest impact on your life.

Your ability to work with others is one of the key skills you will need in order to succeed in college and your life beyond, whether in a career, graduate studies, or the military. Thus, I cannot overemphasize the importance of developing and maintaining a social life while you are in college.

This does not mean that you need to go out and party; it means that you will want to make friends with, work with and for, supervise, and make mistakes with, fellow dorm residents, classmates, and student club members.

STEM students have a reputation for being introverted,[16] but don't be

---

[16] If I have to hear the joke about extroverted engineers "staring at the other guy's shoes" ever again, I'll vomit.

fooled: introverts are not antisocial. It turns out that introverts are actually social people; they just have more discerning taste—well, in my opinion, anyway!

To lead from this, it is also a common misconception that all introverts are shy. In reality, some are, and some are not. If you have read Susan Cain's book *Quiet*, you'll know that introverts need to be around people—just not a lot of people, and not all at once.

Why bring this up in a book about STEM study habits?

First, meeting new friends at college is part of the fun (and part of the drama), and making a new friend everywhere you go is a good habit to start when you leave home for school. You will want friends to grab a bite with, meet for study appointments, and find out what adventures might be happening outside your circle of familiarity from. If you meet students a year or two older than you, they can give you a heads up on which courses to take and which to avoid, or which professor has better lectures and which one tests on details found only in the footnotes. In the same vein, you may find yourself coaching a friend a semester or two behind you in school and thus returning that favor. We learn best when we teach others, so don't miss out on that opportunity!

Second, it will be through relationships that you will establish and build your career. You will need letters of recommendation, a heads up when a new position becomes available, someone to move your résumé to the top of the pile, and perhaps even someone above you to defend your position when layoffs happen.

If you have made it this far into the book, I know you will study hard, work your homework problems frontward and backward, and learn your technical skills. I have no doubt about that. Learning *people*, however, can be much more difficult. With that in mind, let's talk about three different ways in which you can meet new people, make good mistakes in a low-stakes environment, and learn social skills.

Social Person

The friends you meet in college (and shortly thereafter) are friends you will carry with you for a long time. Part of the reason for this is, you are all going through an exciting adventure together, and this has a galvanizing effect. Another part of it is the fact that you only make lifelong friends early in life. When we get older, we still make friends, but they will never be "lifelong" friends. The math does not add up.

My parents were newlyweds when my dad took off for Officer Basic after graduating college. As two South Dakotans eight hundred miles from home and deep in the south, this new life started off a bit lonely—that is, until my dad, a young Second Lieutenant, took my mom to dinner at the invitation of a hard-boiled colonel and his socially graceful wife. My parents were new to the post and area, and this colonel, who was well seasoned and had served around the globe and seen everything one could see in their career, gave my parents a piece of advice that they passed onto me: "If you make one or two good friends everywhere you live, you will be better off than most who stayed in one place with a large group of acquaintances."

We all need friends who understand us, support us, celebrate our victories with us, and wipe the dust and tears off when we crash. I enjoy meeting people, hearing their stories, and learning what makes them excited about life.

I ended up following that old colonel's recipe and making one to two really good friends every time I moved.

Keep it simple to start with: introduce yourself to your dorm neighbors and classmates sitting next to you, ask a classmate who looks as lost as you feel if they would like to study with you, join a club with likeminded souls, and (with rare exception[17]) always accept an invitation if you receive one. Oh, and smile!

---

[17] This being your safety. If you do not feel safe, do not accept. Full stop.

Social Life

We are not meant to live as isolated automatons. Technology (and I am a big fan of much that technology has to offer) has a downside: it makes virtual connections easier than ever, and this means we face the sad outcome of living with more virtual connections than physical connections.

Remember that online friends are not the same as in-person friends. Close your laptop and join the ranks of the humans on campus.

Making friends can be awkward, and you will not get it right every time, but it is worth finding and developing supportive friends. Take inventory of your current friends: do they bring value, and do you bring them value? If they (or you) do not, then it may be time to develop a new circle (or do some self-reflection).

One of the critical factors for success is surrounding yourself with friends who have similar values. You do not need to think the same politically or be in the same degree program, but you should all be focused on learning as much as you can and enjoying the time you have in school. You do not want friends who encourage you to blow off homework or skip class. Kindness and generosity are also two attributes that all my friends that I have kept close have in abundance.

Your social life, then, should reflect your friendship circle. Make plans to meet for lunch, go out on a Friday night, or go train together. As important as studying will be in your college career, study appointments are not a social life: you need to go out and laugh, have fun, see a movie, or attend a concert or game. If you are feeling particularly brave and you live off campus, get your study group together and invite your TA or professor over for dinner. Spaghetti is the easiest meal in the world to make: a pound or two of beef, a jar or two of spaghetti sauce, and a box of spaghetti. Grab a bag of greens for a salad, heat up a loaf of bread in the oven, and you have a meal. No one expects it to be gourmet; they just expect to enjoy the evening.

Building a social life, particularly if you have not done so up to this point, will take a bit of getting used to, but you can do it. Get to know your neighbors and your classmates. These people will make a difference in your life, and at some point, you will have the privilege of returning the favor.

Social Media

By the time this book goes to press, there will likely be a new social media platform that I have not heard of, much less been on.

I am not going to caution you again social media. I will, however, gently caution you about the amount of time you spend on social media,[18] its potential negative mental health consequences, and, of course, the reality of future employers looking at your posts. You already know that everything you put online is captured, so if you have any doubts about a post, wait until the next day, or clear it with a levelheaded friend. You never have to justify, excuse, or apologize for a post you did not make.

One platform I do recommend is LinkedIn. Every professional organization has an account, and much good information can be mined from company profiles, industry news, and individual profile pages. If you are looking for a professional career after graduation, I recommend setting up your profile page so it is consistent with your résumé. Have a friend review it for typos or unintentional inconsistencies, and update your profile every time you update your résumé.

Another opportunity available through LinkedIn is the ability to look ahead to the careers of individuals who are much further along in life. Did they obtain any additional credentialing or licensure? Did they earn an MBA? How long were they at a given company for? How long did they spend in various positions within a company? I recommend sending them a connection request, and if they accept, consider sending a message to introduce yourself. You might even be able to ask about an informational interview (more on that in Chapter 13)!

There are also those who keep up with a personal website, though I

---

[18] Seriously, time yourself. I promise you will be stunned at how much time disappears in mindless scrolling.

personally am seeing this less and less. If you choose to do so, please be diligent in keeping it up to date.

Social Cohesion: Pulling It All Together

Downtime versus Social Time

If you tend more toward the introverted side of the scale, it is easy to burn out or be overwhelmed with the sheer volume of social... everything. There is nothing wrong with taking time to hibernate and recover. As extroverted as I tend to be, I still need my downtime during the day and on the weekends.

One of the hardest lessons I have had to learn is that I cannot accept every invitation sent my way. For me, letting go of FOMO (fear of missing out) was a difficult and necessary habit. I say "habit" because it was always my habit to accept every invitation and then try to figure out later how I could make it all work without faceplanting from exhaustion. I also know that spending too much time alone lends itself to brooding, which does not do me any good, either.

The mix never seems to be the same in a given day or week, so just keep putting yourself out there and refining your proportion of quality alone time to social engagements.

By the way, it should be noted here that I have yet to meet anyone, of any age, in any walk of life, who actually has this figured out. If you do get it figured out, drop me a line and let me know how to do it!

Homesickness

Something we do not talk about as much anymore is homesickness. I suspect this is due, in part, to our technological connectedness. What I do not know is if this technological connectedness is an adequate salve, or if it just masks the symptoms.

Homesickness results from a lack of feeling like one belongs in their current living environment. How do we start to feel like we belong? Let me first put your mind at ease: no one shows up to a brand-new school or event and feels like they have it nailed down immediately. We all feel awkward, and a bit lonesome. There will always be those individuals who appear to have it figured out, with their brimming social calendars and exotic social media posts. There may be many people around you who seem to be ahead of the game (whatever the game may be). The good news is that you know enough by now to know that this level of exceptionalism rarely (if ever) exists. People like to put their best foot forward, or even (as is all too common nowadays) create an entire life for public consumption.

What you are charged with figuring out is you. We start making friends, we get to know and work with those around us, we find our fellow likeminded students and grab a meal or coffee together, we plan camping trips or hikes, study together, play video games together, go to sports games together... the list goes on. Sitting alone in our dorm rooms, stuck on social media or FaceTime, will not actually resolve the problem of homesickness. Building up and participating in the community around you, however, will.

Balancing Socializing with Work

With all this socializing, how can anyone expect to get any work done?

Socializing is fun and, once you get the hang of it, might seem much more appealing than focusing on your homework (see Chapter 5 for all things procrastination). If this is you, first remember your purpose for being at this school: to earn your degree, preferably in a reasonable amount of time. Second, remember you can double dip studying and socializing to the extent that it is useful (see Chapter 3 for all things exams). You can, and should, make friends within your extracurricular activities, like clubs or student chapters, and, of course, your roommates or dormmates (though I understand roommates can be a bit dicey, as you don't always get to choose who you live with when assigned a room).

Studying in small groups (i.e., less than six people) is a wonderful way to learn the material and keep yourself connected to the rest of humanity. Ensure your homework and exam calendars are up to date, schedule your free time and time with friends, and do not worry about getting it right all the time. You won't. Neither will anyone else.

Recommended Reading

- *Quiet: The Power of Introverts in a World That Can't Stop Talking* by Susan Cain.
- *How to Win Friends and Influence People* by Dale Carnegie.
- *Never Eat Alone: And Other Secrets to Success, One Relationship at a Time* by Keith Ferrazzi and Tahl Raz.

Strategies for Success: Get Social

- Make one to two really good friends everywhere you live.
- Build your friendship circle with smart people who are kind and generous.
- You cannot work all the time, so go out and have fun, and be extremely circumspect about what part of your fun you post on your social media accounts.
- Build and maintain your LinkedIn profile in keeping with your résumé.
- Use LinkedIn to see what career paths and options there are. Reach out to someone who does what you think you might like to do and interview them.
- Homesickness and loneliness are real. The best antidote to both is to get out and introduce yourself. Start making friends.

11

Health

Almost everything will work again if you unplug it for a few minutes,
including you.
—ANNE LAMOT

WHAT DOES PHYSICAL, MENTAL, AND spiritual health have to do with finishing homework assignments and passing exams as a STEM student? Actually, everything. Our brain is an incredibly precious tool, and if we do not take care of that tool in all three capacities, it simply will not have the utility that it could and should.

In this chapter, we will start with physical health, as the habits pertaining to this area of our wellbeing are some of the easiest to observe and adjust. After we have covered physical health, we will move onto mental and spiritual health.

Remember, you cannot manage what you do not measure, so pick a habit or two and start tracking in your calendar. When you have those down, move onto the next one. Start with whichever habit will give you the biggest impact for the least effort (the Pareto principle).

Physical Health

There are various good physical health habits you can (and should) adopt, and the specific habits listed in this chapter directly support your brain. As a STEM student, you are creative, and for you to keep this attribute, you need an alert, focused, well rested, properly fueled, hydrated brain.

Sleep

Sleep is one of the most, if not *the* most, dramatically underrated health necessities of them all. We need to spend about a third of our twenty-four-hour day sleeping.

Years ago, I wanted nothing more than to decrease my need for sleep like some of the notables—Arnold Schwarzenegger, Margaret Thatcher, and General Stanley McChrystal, to name a few. Approximately four to five percent of the human population only needs about four to five hours of sleep, and I desperately wanted to be in that group. Just think how much I could accomplish every day by cutting my sleeping hours in half! Thus, I tried conducting an experiment where I went to bed at the same time every night but shaved fifteen minutes off my wakeup time every morning.[19]

Sure enough, by the end of the week, I was so exhausted that my speech was slurring, I could not concentrate, and when I finally arrived home after work, I sat down and promptly fell asleep for three hours. Any time I imagined I was going to save by shaving off my night of sleep was lost, and certainly no benefits were accrued.

When I am exhausted, the first thing to go is my creativity, and next, I lose focus and motivation. Finally, in a desperate effort to stay awake, I start eating too much sugar. In the end, I have decided it's just better for me to just get the sleep I need and make the most of the rest of the day that I have

---

[19] There were many issues with my experiment, of course, sample size being one and my lack of understanding of sleep cycles being another.

available to me.

Most humans need six to eight hours of sleep every night (and no, in all likelihood, you are not in the four to five percent of the population that this doesn't apply to). You will need to try out what works best for you in terms of your needed quality and quantity of sleep. Over time (and by keeping an eye on my tracker), I have discovered that of the seven hours of sleep I need, at least one hour needs to be deep sleep and at least one hour needs to be REM (more of both is better, but at a minimum). Taken together with the amount of time I'm awake, this adds up to eight and a half hours on the clock. This means I am in bed by 9PM and up just after 5AM. With that amount and combination of sleep, I feel pretty darn good throughout the day: I can accomplish what I need to in terms of creativity, productivity, and effectiveness without relying on sugar or too much caffeine.

I am also a big fan of taking naps. Thor Heyerdahl used to take a ten-minute nap every afternoon. He would hold a spoon, and when the spoon dropped, it would wake him up and tell him it was time to finish his nap. I personally find twenty minutes to do the trick. I use earplugs, look at my watch (envisioning twenty minutes), and then fall asleep. As much as I love coffee, this twenty-minute nap is so much more reinvigorating that I would recommend it to anyone needing a little pick-me-up in the afternoon. If it keeps you from eating junk or sugary foods, so much the better!

Food

In some ways, the food section is the simplest here, yet it feels like the most complicated. I think this is because we *want* to make it more complicated than it is.

My rules for food are as follows: first, if I cannot pronounce the ingredients, I do not eat them. This has the wonderful impact of eliminating most (if not all) highly processed food. I have no interest in putting chemicals into my system that I cannot account for, and by keeping processed foods at a minimum, I know I am making a straightforward decision that will have a lifelong impact.

What I will tell you is that the hardest part about food is portions. Portion control is a huge challenge, particularly when you are in college, because if you are eating at a buffet where you pay one price no matter the portion, of course you're going to want to get the most food for your money. Resist this temptation, my friends. Use a smaller plate so it looks full if you need to, but do not overstuff yourself just to ensure you are getting your money's worth. The cost will be your life expectancy. Do not go back for seconds. If the options are to be a little overstuffed or a little hungry at the end of your meal, you are better off being a little hungry, trust me.

Two more tricks for portion control:

First, always leave something on your plate. I know this goes against what every one of us who grew up with parents told us ("Clean your plate!"), but if you want to maintain portion control, then either put a smaller helping on your plate in the first place or leave some there at the end.

Second, when you are eating out, as you are served, ask for a to-go box and immediately cut your meal in half before you start eating. The meal is usually delicious, and given the choice, I would rather eat it start to finish, even though I know it is usually far too large a portion for me to reasonably consume. But put half in a to-go box, and you'll have leftovers for the next day—leftovers that will support efficient living habits and will mean you don't have to prep another meal. Bonus!

There are obvious exceptions to the "portion control" rule: if you are a student athlete, your caloric intake needs to be greater than those of us who are predominantly sedentary. Work with your coach or team nutritionist to determine what type of fuel your body needs for optimum performance.

The other exception to reining in your caloric intake, of course, is anyone suffering from an eating disorder. If this is you, please work with your doctor or counselor to ensure you are maintaining healthful eating habits.

Another problem for me is that I love pizza. This means any student organization that offers free pizza will immediately pique my interests. But as perfect as pizza is, too much of anything is not a good thing, and depending on how cheap the pizza is, it can be a salt bomb or a chemical festival, neither of which is good for your brain. Pick one meal a week to be your pizza [or your other junk food of choice] indulgence, and bypass the rest. Your brain and body will thank you for it.

A longtime student (now professor) I know observes, "Quite often, the taste of food is inversely proportional to the cost of that food," meaning the cheaper it is, the better it tastes. And in college, we do not often have the luxury of high-end food choices. You won't go wrong with the basics, like making yourself a sandwich for lunch, oatmeal or eggs for breakfast, and something straightforward for dinner (like a lasagna or hamburger), but I do encourage you to start putting salads in your diet if you are not doing so already.

Before you roll your eyes, understand that it took me a long time to consistently put salads in my meal plans, okay? I eventually figured out that if I put steak tips on top of my salad, I enjoy them more. Do whatever you need to do in order to get your five a day, because the facts are, we need to eat greens every day, preferably in at least two of our three meals.

Sugar is another one that gets people, including me, but do your best to cut that out of your diet, nonetheless. There are a host of reasons why this is only ever a good decision, and you can find them all through a quick search on the Internet. Suffice to say, sugar will interfere with your ability to concentrate, which will interfere with your ability to do well in school— and to me, at least, that's reason enough to keep it to a minimum. See the section on refined sugar later in this chapter if you're still not convinced.

Exercise

Exercise seems to be one of the first major cuts from a busy person's daily routine (after sleep), yet it is just as vital as rest. Exercise keeps blood flowing, introduces fresh oxygen into our system, gives us a much-needed mental break during the day, and helps us sleep at night. It also helps us maintain a healthy weight and feel good about ourselves.

Forms of exercise can vary, from a brisk walk to endurance competitions, and everything in between.

Some of the chapters in your life will be busier than others, which means you need to be adaptable when it comes to keeping an exercise regimen. At a minimum, you need thirty minutes every day.

When I was really pressed for time, one of my favorite mechanisms for getting a bit of exercise was to schedule walking meetings. We had a loop by campus that was about half an hour around, and when I had to meet with somebody, we would go walk and talk at the same time and accomplish both goals: the meeting and our daily exercise.

If you can commute to school on foot or by bicycle, I highly recommend doing so. The fresh air at the beginning and end of your day will do you wonders, and it'll also mean you won't ever need to schedule a separate workout time if you do not want to.

If these ideas aren't something you can do, there are some basics that you can incorporate every single day: sit-ups, pushups, burpees, squats, lunges, and short runs. There are also more involved options, like strength training programs, dance classes, yoga, spin classes... you name it. If you live near a state or national park, you can go hiking, climbing, kayaking, or trail running. Regardless of what you choose, take the time to find a variety of activities that you enjoy while in school and that are suited to different seasons. Then, when you are out of school, you will have a battery of options at your disposal (and the means to invest in your new hobby)!

Water

Drinking water matters a great deal for your brain and cognitive processes. I drink sixty-four ounces (a gallon) of water every day.

Yes, a gallon of water every single day.

At first, that sounds like a lot, but here is how I break it down: I have a sixteen-ounce refillable water bottle that I carry with me. At a minimum, I drink one bottle before 10AM, the second before lunch, the third before 2PM, and the last by about 6PM. Any additional water is a bonus.

By breaking the goal down into small pieces like this, I don't have to worry about drinking a gallon all at once, and I feel great throughout the day. Being hydrated makes both thinking and working out easier, decreases your craving for sugar and artificial stimulants, keeps the brain fog away, dispels headaches, and helps you feel lively. I do *not* recommend setting a goal of sixty-four ounces every day and then leaving the task until last thing in the evening.

Student athletes will likely need more water than this sixty-four-ounce-a-day recommendation, so if you fall under this category, follow the recommendations of your coach or trainer.

Personal Hygiene

Personal hygiene is something no one talks to you about, but everyone notices.

In this culture, it is expected that you shower or bathe at least once a day. Body odor is particularly offensive, but is something no one really wants to point out to the offender, so here is the blunt conversation nobody wants to have with you: the smell of body odor is the decomposing bacteria on your body and in your clothes. The worst part is, if you let your hygiene slide, you may not even notice just how bad it is. Olfactory fatigue (or odor fatigue) is a temporary inability to distinguish a particular odor after prolonged exposure to it. This means *you* won't notice your smell, but your roommates, classmates, and friends certainly will.

The good news is that it is not permanent, and is easily fixed.

For one, you need to change your clothes every day, including your underwear. Depending on how often you are willing to do laundry, this means you need to have at least that many days' worth of undergarments (i.e., underwear, undershirts (if you wear them), and so on). For example, if you only wash your clothes every two weeks, then you need at least fourteen different pairs of undergarments. If you wash clothes every week, then you can get away with seven or eight pairs (you will probably want to be wearing a set when you do your laundry).

Second, when you shower, you need to use soap and scrub down all the nooks and crannies. Men and women alike should also wash their faces every night to get all the oil and dirt they've accrued throughout the day off. Use a very simple and light moisturizer for your skin at night (yes, men, I am still talking to you, too). It should be unscented, and it does not need to be expensive, but for heaven's sake, do not use hand lotion or body lotion. Both men and women should also put on a moisturizer in the morning that has an SPF of at least 15 built into it. Trust me. When you are in your forties and getting your skin checked, you do not want to have bits and pieces snipped or frozen off.

In addition to showering every day, you need to wear deodorant or antiperspirant.

Again: every day, you need to shower (and use soap in all the skin folds), wear deodorant, and put on clean undergarments. If you wear synthetic fibers (like fleeces or polypropylene shirts), you will need to wear a fresh one every day. Your roommates, classmates, and date will thank you.

Now, let's talk teeth. Go to the mirror and smile. Besides bad breath (or halitosis), if you have yellow teeth with material caked in between and dark red rims around your gums, you have gum disease. If you have white teeth and pink gums, good job: you are probably brushing and flossing sufficiently. At a minimum, visit the dentist once a year (preferably twice) for cleaning and a checkup. If you chew or smoke tobacco, these checkups are especially important so you can check for precancerous lesions. If you are too tired to both brush and floss at night, always choose to just floss.

Men, if you are able to grow a beard, by all means do so, but keep it tidy. If you are on your way to a beard but are not quite there yet, then I recommend a clean-shaven face for your interviews.

When you comb your hair, remember to comb the back and not just the front. Keep your nails trimmed and your clothes clean.

Mental Health

When we talk about mental health, we are talking about your ability to focus and concentrate on the work needing to be done, your overall attitude toward school and your social life, and your ability to rest and recover. If you are distracted by unfinished work or items needing to be put on your to-do list, stop what you are doing and write them all down. If your attitude is one of constant frustration, despair, or irritation, then you are probably overloaded. Go back through your calendar and start cutting items from it. You need to take real time off. Go look back at the quote at the beginning of this chapter to see why!

If you are not someone who generally gets physically ill and rarely needs a sick day, then by all means, take a mental health day. I am not someone who generally gets physically sick, so whether I'm in the professional world or in school, if I simply need a day off, then I take a mental health day—a "Scyller Day", as I call it. During this day, I do not worry about checking emails, keeping up with social media, or even really answering my phone or text messages. When I've reached the point of maxing out, then I know it is time for me to cut back and spend time focusing on activities that rejuvenate me.

We will talk about hobbies and creative energy in the next section (under "Spiritual Health"), but for now, know that it is so important that you keep track of and note your mental health as you increase your responsibilities both academically and socially.

Stress... and More Stress

Believe it or not, stress (and more stress) is a normal part of college. Just as you stress muscles out in order to strengthen them and make them grow, being under a certain amount of mental stress is actually good for your cognitive growth (i.e., your toughness and resilience). For most of my life, I have seen stress as a companion: stress gets me up and off the couch, gets my work done, ensures I put downtime on the calendar, and ensures I finish what I start. Stress is not necessarily a bad thing. Enter an environment that is completely devoid of stress, and you will atrophy.[20] Accordingly, when you go into the professional world in any capacity, expect to be moderately stressed.

Stress is too much when you are losing sleep, your eye is twitching, or you are pulling your hair out. Not enough stress, on the other hand, is when you are laying on the sofa mindlessly binge-watching some series you cannot even name and eating food that's been delivered to your house. If the biggest stressor in your life is getting off the couch to answer the door and grab the food being delivered, you do not have enough of it in your life.

One of the reasons why I experienced mild insomnia when I first went back to school was, I always started homework assignments too close to the deadline. This is the kind of stress you'll do well to avoid. You do not need to finish every homework assignment the moment it's given to you—that will not realistically work with a busy schedule and with how long some assignments take—but if you immediately review what you have coming up and make a plan for where it will fit into your schedule, you will be able to sleep at night.

See the previous section on physical health (specifically exercise) for alleviating stress. The endorphins you experience post-workout are a terrific way to alleviate stress (and more stress).

Routines

If you look up the difference between routines and habits online, you will find as many opinions as there are answers.

Here is the distinction that I personally make: routines are set in time,

---

[20] Please note, I am reserving the term "anxiety" for the diagnosed condition. Stress, even though it can feel heavy at times, is not debilitating as anxiety can be. If you find that you are unable to function because of your stress, this is in fact anxiety. Please seek help. Most universities have counselors on staff or access to mental health professionals who can help you.

while habits are tied to practices. For example, a healthy *habit* is to wash your hands every time you use the toilet. *When* you use the toilet is up to you, so it's not an activity that's dictated by time (and it therefore isn't a routine). A *routine* is tied to a specific time, like your morning routine or your evening routine (which is when you know ahead of time exactly what you will do when you wake up and exactly what you will do before you go to bed).

Through routines, you eliminate much consternation and frustration, and you take away the wasted time and effort involved in redeciding decisions that quite honestly could be made once and be done. A periodic review, of course, is recommended, but at the least, you have eliminated standing in front of your closet or mirror in the morning for half an hour if you have already pre-decided how the first hour of your day is going to go.

On the topic of routines: if your intention was to be in bed by a certain hour every night but you have not taken the time to line up everything you wanted to get done by then, then—news flash—you can't decide at 8:57PM to be in bed by 9PM *and* to still do twenty-five minutes' worth of activities (and expect to still get into bed when you want to).

To prevent this from happening, here are my morning and evening routines. If they work for you, wonderful, though I still strongly encourage you to try out and read up on different routines so that you find one that fits with your needs and your schedule.

My Morning Routine

My morning routine has not changed much since I was in college. For my first bachelor's degree, though, I was a rower, so my mornings were dictated by my workout schedule. Other than in those scenarios, this is how my mornings go:

My coffee pot is already full because I program it the night before to brew ten minutes before I wake up (which is around 5AM). After my morning ablutions, I grab my coffee and head to my office or living room and do my

Bible study, my prayer time, and writing. Depending on my overall schedule, I may work out in the morning, or, if I have the option, I will save my workout for later in the day as a reward for getting my work done. If something absolutely must get done, I will do it in the morning. There is so much about our day that we cannot predict, but if I have everything that needs to be done complete by 7:30AM, then the rest is a bonus!

It is a herculean effort, but I also try to stay off my phone until my morning routine is complete.

My Night Routine

It takes me a bit to wind down for bed, so I end up starting about an hour out. I am generally in bed by 9PM, so starting at 8PM, I shut down my laptop and phone or tablet and start getting ready for bed. For me, this includes sweeping through the house and tidying up a little bit, putting the last of the dirty dishes in the dishwasher or washing them by hand, putting any dirty clothes or towels in the hamper, straightening up the living room, doing a quick check on the doors and windows for safety, and making sure my coffee pot is prepped and ready to go for the next day. If I have an early start, I will lay out my clothes and anything else I'll need before I head out. I will also double check my calendar to make sure I am not surprised by a commitment or event.

I take my vitamins (multivitamin, probiotic, and cod liver oil) at night because I include one that helps with muscle cramping at night (magnesium). This is not necessarily the best time to take my supplements, but I know it means I take them consistently, so the tradeoff is worth it to me.

I wash and moisturize my face, brush and floss my teeth, and sometimes do a little writing or reading. Then, I'm into bed, excited for the next day to start!

When It's Too Much

During my career as a specialty pharmaceutical sales representative, I spent a year working with psychiatrists in the mental health field. The patients we served suffered from severe forms of debilitating mental illness. Some voluntarily admitted themselves, others were involuntarily admitted, and many had to be medicated several times a day.

To prepare for this role, I reviewed the *DSM IV*, the gold standard for diagnosing mental health disorders at the time. Reading through the pages of illnesses and diagnoses, I slowly realized that I had half the diagnosis for the mental illness my patients were suffering from, and my family and friends had the other half. So, why weren't we all on medication?

It turns out everybody has what I would call "filaments" of these issues.

While visiting one of the psychiatrists, I asked him about this observation. He said that you seek help when (and only when) your issues are preventing you from interacting with society in a functional way. If you are not able to go about your business (work, socialization, and downtime) as a result of your symptoms, then you should seek help. While this is not a formal definition for when to seek help, I think it is a good starting point.

Mental health concerns, diseases, and disorders generally present themselves when people reach their early twenties, which means that around the same time that students leave home and navigate a new life with newfound freedom, they may also be experiencing life altering circumstances in their mental headspace.

If you suspect that you or a friend may be going through something that is more than your run-of-the-mill college stress, I recommend that you reach out to a mental health professional on your campus. There is no shame in seeking help and pursuing a diagnosis and treatment so that you can be at your very best and accomplish all that you are capable of in this lifetime. You may need something as basic as counseling, or a much more involved regimen.

As a starting point, ensure that you are getting your sleep (quantity *and* quality), not loading up on sugary or chemical-laden drinks, and not self-medicating with alcohol and marijuana (or any other drug). Also ensure that you are getting your exercise, socializing with friends and classmates, and consistently knocking items off your to-do list (even if it is just folding your laundry or putting your class binder in order). These are all basic recommendations that will contribute to good mental health.

Quitting

The word "quitting" carries with it a very negative connotation, but the truth is, there are times when it's appropriate to walk away, just like there are times when you simply must dig in and finish the thing. So, how do you tell which is the right course of action?

First, let's talk about when you should walk away.

I once had the experience of holding an office in a department club in which the president started using funds for questionable activities. In my view (as the treasurer), he was making decisions without the consent of the board and, ultimately, for his personal benefit. I reported this to the club advisor, gave him my notes, and turned in my resignation. Yes, I'd committed to holding that position for the entire year, but I was not about to participate in something that I believed to be unethical, and certainly not in a group with a leader whose values didn't align with mine.

Another example would be when I graduated from high school and participated in the national competition for Lincoln Douglas debate. As I have mentioned previously, I loved debate in high school—it was my entire life—and one time, on my way to a debate meeting in college, I ran into the men's rowing team, who were recruiting for the women's team. Then and there, I quit my interest in debate and signed up to be a rower. I had fulfilled all four years in high school, and now, it was time to try something new.

Just because you did something in high school and were successful in it does not mean you have to carry it forward into college.[21]

---

[21] Of course, the exception here is if you receive scholarship money for this. In that case, you need to account for that in your decision-making process.

If you are considering quitting a role with responsibilities that you agreed in advance to take on, consider finishing out the term (or the semester, or the school year) to put a bow on it. If it is voluntary and you have not really committed in any way other than by showing interest (and now your homework and exam schedule are pressing down on you), give it the heave ho.

When do you *not* quit? Simply, you do not quit just because something is hard. You do not quit because you're frustrated, cold, and hungry. Every project, goal, endeavor, and dream worth pursuing will have a point between seventy-five and eighty percent of the way through where you are just sick and tired of it and want to quit (for me, my master's thesis, my doctoral dissertation, every half marathon, putting my laundry away, and writing this book, to name a few). Whether it's a semester, a term paper, a race, or a senior capstone project—it does not matter—there *will* come a point when you simply do not want to do it anymore.

That's the bad news. The good news is, that point comes when you are almost at the finish line. If you have pushed through this feeling before, you know exactly what I am talking about, and there is no victory as sweet as when you have forced yourself to see the thing through and get it done.

Do not quit your class, project, or degree program. Yes, it will be hard, but pushing through the hard is what will let you stand up proudly on graduation day, knowing that even though you maybe had to crawl across that finish line, you crossed it anyway.

Spiritual Health

If physical health pertains to our bodies and mental health pertains to our minds, where does spiritual health fit in?

I am a Christian, and view the world and my place in it through my faith. I believe that, as human beings, we are meant to be in community with one another. I believe that we all have a purpose in this life and a place in this world.

Whether you profess Christianity or not, I expect you realize that there is more to this life than simply working, working out, eating, and sleeping. Hence, this section is here to give you a broad guideline regarding your spiritual health based on the experience I have had visiting students and knowing that we are intended to be well rounded individuals who are part of a greater body.

Religious Practices

Growing up, I went to church and midweek services, but when I left for college, I had a tough time making it to church on Sunday mornings. As a rower, we had practice at 5:15AM every weekday morning and usually a regatta on Saturday mornings, so that meant my only time to sleep in was on Sundays. This simple fact had a devastating effect on my church attendance. I did find after some time, however, that a late afternoon service was offered, and many college students took advantage of this option.

If you grew up attending church or religious services, I strongly encourage you to stick with that throughout your college years. It will serve as a strong reminder that just because you are living in the pressure cooker of college does not mean that that is all there is to this life.

Many members of your church family will serve as extended family, which is wonderful, particularly when it comes to hearty meals! My church family always made sure that I was well fed, and the cooking was unbeatable.

I can also assure you that my prayer life was never so active as it was during finals week.

There are numerous nondenominational student organizations on campus, and you might want to try those out. This will also have the added bonus of introducing you to students in different disciplines and degree programs, which will help to broaden your view of university life.

Colleges frequently have religious services at times that are more amenable to a college life and sleep schedule, so keep those spiritual practices going strong.

Creative Outlet

Hobbies, in many ways, are a luxury: they require time, money, and thoughtfulness. For most in the world, any one of these three constituents is scarce or nonexistent. Yet in more privileged worlds, we experience such a paucity of human culture that I am no longer convinced we are well off. Surfeit does not mean satisfied. From having any number of streaming options for shows, to more products in a grocery store than could ever be imagined,[22] we have access to more "things" than ever before, and yet are still found lacking.

In this world of "rapidification", hobbies matter more than ever for our wellbeing. We need not do these activities because we have to be the best, fastest, or most creative (despite what we may see on social media feeds); we do, however, need hobbies so we can slow down and appreciate every one of our senses and the beauty they absorb.

This summer, I planted a container garden. Some may consider sunflowers to be weeds, but I do not care one bit. In spite of the package instructions (and because they were on sale), I started my seeds in mid-July (that's the thing about weeds: they will grow regardless), and let me tell you, I *love* the smell of soil after watering my lovelies.

Shortly after I wrote this section, I came across a book on geraniums in the used bookstore. Who knew there are more than four hundred varieties of pelargoniums? Interest piqued, one of my many Facebook club memberships now includes Pelargoniums International.

In the mornings, I go out with my coffee cup and I pull weeds, water each container, encourage my little plants, and listen to the sound the

---

[22] From nearly nine thousand items in 1975 to over fifty thousand items in 2023. Yes, really. And that doesn't include online megasites.

breeze makes in each plant. Every plant sounds different.[23]

My sunflower seeds sprouted, and despite an evil neighborhood cat marking his (my) territory on my soil, they grew, and their sunny faces now greet me when I return home.

Then, as the school year started, I decided I wanted to learn how to make bread from scratch. My first loaf (a potato flour bread) was a little dense. I still have much to learn about proofing the yeast, kneading time, rising, and punching, and yet I beamed broadly at the finished product. One of Martha's[24] rules is to "make it beautiful", so before baking, I brushed the top of the loaf with coconut oil and then sprinkled salt and savory spices over the top. Then, once it was baked, I enjoyed slice after slice—just toasted with butter and homemade jam.

Is there anything as decadent as a slice of homemade bread, lightly toasted, with sweet cream butter and jam?

So many of my friends and loved ones participate in hobbies— woodworking, canning, knitting, painting—that it gives me pause. Why do these hobbies? Certainly not for financial gain, international renown, or résumé-building. Simply, hobbies humanize. Without hobbies, my appreciation for true artists could only be anemic, if it existed at all. How would I understand the care and practice and refinement that went into such artistry if I had not first tried my hand at a smaller fractal?

Hobbies equilibrate. Humming while I weed, repeating the rhythm of kneading, and generally getting lost in my task, gives my brain a break. Drama moves so far to the side of my thinking that I no longer worry, and my shoulders drop about six inches.

Hobbies satisfy. Grades, performance assessments, and rubrics brook no currency with hobbies. Smell, sound, sight, taste, and haptic sensation count. Puttering along with my plants or bringing a new loaf of bread to life satisfies my atrophied senses.

Set aside the gerbil wheel newsfeed scroll and get absorbed in something beautiful. You can read about the "seven most important morning habits of

---

[23] The sounds of air running through tall grasses are my favorite, in case you were wondering.
[24] *The* Martha Stewart, obviously.

successful CEOs" tomorrow. Or never. Either works.

Today, cultivate your hobby.

Volunteering

Volunteering is possibly the best way to realize that the work you are doing will have a greater impact on the world around you, and that in fact, life is not all about you.

It is so easy to start only thinking about yourself when you are focusing on the habits you are building to create a successful and effective life, but living in a community means your life is (at least partially) about your place in said community. You have a role and a responsibility to take care of yourself and take accountability for your behavior, and you also have a responsibility to those around you.

Volunteering has a way of using the social interactions we'd normally avoid to sand down our own rough edges.

You can volunteer weekly, monthly, or annually, and you can do so in a way that fits with your current skillset (or helps you develop a new one). One of my favorite ways to volunteer is to read to children in elementary schools. While I was never inclined to babysit or be an elementary school teacher, I very much believe in the importance of reading and enjoying reading (as you'll know if you read Chapter 8), and if I can help these little ones learn to love reading simply through a good story, then I have had a success for the day.

You can alternatively volunteer for larger projects like Habitat For Humanity, or you can start your own program, like a student I knew during my doctoral program. He dedicated one night a year to helping make sure all the homeless had a place to sleep and get out of the cold, and he was remarkably successful in getting both the university and the greater community to work together. When he graduated, he had already made an incredible impact on our community.

"Do not be afraid to start small," said Willie Nelson. I agree with this advice, and accordingly encourage you to start small in your plans (but big in your dreams). As you grow in your competence and skills, then by all means, grow the size of your project. Until then, enjoy the fact that when you volunteer, you can take a break from thinking about yourself entirely, and focus on someone else for a while. This is a great relief.

Avoid the Poisons

Here are the big three: tobacco, drugs, and alcohol. Zig Ziglar talks about these three in several of his speeches, so I cannot take credit for lining them up and coining them "poisons", but I can say that I agree with him. Let me be clear: just because someone *can* enjoy a poison, does not mean that someone *should* enjoy a poison.

Consider the fact that our brains do not finish forming until we are about twenty-six years old. This means that introducing the poisons to our brains in college will probably have real and lasting consequences on every area of our health and wellbeing, potentially for the rest of our lives.

States are now moving to prohibit tobacco use by anyone under twenty-one (just like alcohol), and for good reason: liver disease is on the rise in younger people as a result of drinking, as is heart disease (as a result of smoking and vaping). And, of course, drug use is pervasive on campuses.

I humbly suggest taking stock of whether you really want to be ingesting any of it at all. If you are in school to learn and graduate with a STEM degree, give considerable thought to whether you want to set back all your hard work and the good habits you have been building through drugs, alcohol, or tobacco. There really is no upside to these three things—only downsides. Besides, I can tell you from personal experience that trying to do calculus problems while enjoying a glass of wine does not work.

If you want to get out and set the world on fire with your genius (and I hope you do!), you will not want to be slowed down by any of these three. You will never hear somebody who is successful stand up and say they owe their success to all the marijuana they smoked in school, but you *will* see people who could have gone on to do great things be undone by these habits (which they often form at an early age).

Stay away from the poisons. They just will not do you any favors.

Two other poisons to consider are screen time and refined sugar.

Screen Time

I cleared out my flatscreen years ago and have not looked back, the primary reason for this being the fact that passively watching television shows or movies is not restorative. My brain never comes away better, more refreshed, or more energized after staring at a screen for hours on end, and that's a fact.

There are two occasions when I allow myself to have screen time: when I watch shows (or parts of shows) while sitting down to eat lunch or fold laundry (for about ten minutes, and then I am back up and on my mission), and when I am on the treadmill.

Based on an entirely non-scientific, sample-of-one study conducted on and by myself, it strikes me that screen size is directly proportional to the amount of time we are willing to spend in front of it.

As a testament to the dangers of screen time, look no further than the online video games that draw in players who do not come up for food or life for days on end. In college, I had a friend down the hallway who became addicted to online video games, and there were three interesting aspects to this. First, it was in the late nineties, when we really did not understand how addictive screen time could be for some people. Second, she was a woman (screen or gaming addiction happens more frequently to men than women, but she was very much in that space). Third, she actually changed shape and color before our eyes. There was a month-long run where she never left her dorm room, kept the lights off at all times, stayed up for stretches longer than twenty-four hours at a time, and only ate junk food out of bags from the vending machine. She did not bathe or shower during this run. She put on considerable weight just in those few weeks, the lack of sun made her translucently pale, and the junk food stained her fingers and teeth.

Eventually, an intervention had to be made because her health (and her roommate's sanity) was at stake.

This is, of course, an extreme example of screen addiction, but then I suppose every addiction ends up being extreme. My point is, in limiting your screen time (no matter how interesting or fun watching TV shows or playing games seems), you will be serving yourself well. When you're binge-watching episodes of a TV show, doing movie marathons, or playing online games, this is time that must be spent in moderation or, better yet, not at all. You will be better served interacting with people, reading books, or listening to good music.

So, when is screen time acceptable? That is up to you. As I said before, I generally reserve it for my time on the erg, treadmill, or stationary bike, and will also sometimes watch parts of shows while eating lunch or dinner. The bottom line, however, is that screen time is neither refreshing nor restorative for your soul. On the contrary, just like the big three listed earlier, it can be abused and addictive.

Considering it has minimal to no upside and considerable downside, decide now how much time you are going to lose (yes, lose, not spend) every week to screen time, and mark it on your calendar. Also note the opportunity cost, which is what you will give up doing in favor of screen time (i.e., visiting with friends or family, listening to or playing an instrument, reading books, outdoor time, hobbies, homework or studying for exams, or anything else that is of actual value).

If this still is not convincing, look at the amount of screen time successful people have: do a quick search for the number of hours spent in front of the TV compared to income level. The more hours of mind-numbing shows watched, chances are, the lower the income level, and this is because we have limited time, energy, and creativity while we're on this planet. How we choose to spend this time matters. What we put in our brain matters.

Refined Sugar

The last poison: refined sugar.

My friends, everything we've discussed so far proved easy for me to minimize or not engage with in the first place. Sugar, on the other hand, is an entirely different story. In his memoir *Life* (I listened to the audiobook version and highly recommend it), Keith Richards describes going through heroin withdrawals three times, and this process was what my sugar withdrawals felt like: headaches, nausea, exhaustion, and irritability.

Okay, so I might be exaggerating slightly, but when you take a week off from sugar, write me and let me know how it went. I, personally, found it shocking.

Part of the reason why it is so difficult to give sugar up is because much of our food is already dosed with the stuff (and most of it is in the stuff you'd assume it *wouldn't* be in). I went through our cupboards and found virtually nothing untouched: ketchup, mayonnaise, bread, breakfast cereal and granola, salad dressings, crackers, spice mixes, spaghetti sauce, tortillas, dried fruit... and that's just to name a few.[25]

As a college student, I do not expect you to gut your pantry and start over with products with no added sugar—that can be quite pricey—but what I do strongly recommend is this:

- Every time you buy to resupply, carefully read the ingredients. Cut out fructose, sucrose... anything with an "ose" at the end. Lactose and glucose are the exceptions; these are naturally occurring sugars.
- If you need something sweet, eat one to three dates. They are sweet and high in fiber, antioxidants, and manganese, which can help regulate blood sugar.
- Remember, fresh food is located around the edges of the grocery store. Lightly to heavily processed foods are in the middle.
- Stay away from soda of any kind. And no, sugar-free is not any better for you than regular.

---

[25] Yes, the manufacturers actually added sugar to *dried fruit*.

- Snack on nuts, not what you find in the vending machine. Watch out for the different roasts, like honey roasted, which can be loaded with sugar.

Why go to all this effort to give up sugar? The obvious answer, of course, is weight gain. Sugar adds calories, not nutrition. Plus, the long-term health consequences of consuming too much sugar are as preventable as they are deadly. Besides weight gain, some issues that can arise due to too much sugar intake are high blood sugar, high triglycerides, high blood pressure, and inflammation.

These may sound like issues which will arise decades down the road, but unfortunately, younger and younger people are being diagnosed and put on medications to solve a problem that a cleaner eating habit would have fixed. Besides, don't you want to help, not harm, your future self?

Sugar can also negatively affect your sleep and your ability to concentrate. Trying to sit down and study while your brain is leaping from one thought to the next because it is high on sugar will not get you the results you want from your efforts.

Plus, your skin. Believe it or not, our skin (our largest organ) is deeply affected by our sugar intake. Too much sugar leads to premature aging and can (because it is inflammatory) exacerbate acne.

Considering all the work you are putting into your degree, the biggest insult of all for a high-sugar diet is this: cognitive decline. High-sugar diets are associated with impaired memory and increased risk of dementia, Alzheimer's disease, and stroke. You are working way too hard to develop your brain right now for you to sabotage your future self in this way. If you cut most of the sugar you currently consume out of your eating habits for a week and find that you have become positively miserable, this means you had too much sugar in your fuel.

Final Thoughts on Your Health

This chapter is a lot, I know. We covered everything from how much and what kind of sleep we need to get, to the fuel we put in our bodies, to the poisons we ought to avoid. We covered how we should spend our mental energy, and how to feasibly engage in spiritual practices.

One thing's for sure: all three legs of the health stool directly impact your ability to learn, process, retain, and make use of the information you take in every day during class, homework, and study.

In school, I wanted to optimize everything. I wanted less downtime, increased study time, less reviewing, and more innate knowledge. Less "wasted" anything and more "effective" everything. If you feel this way, too, I get it, and I am with you on it. Please take note of the fact that your current habits may be undoing your lifetime's practices, and alter your habits accordingly, as per the recommendations in this chapter.

Real change can take time, and it certainly takes energy, commitment, and discipline, so if you do not get it (whatever your "it" is) right away, do not give up. If the results are not immediately apparent (and most will not be), do not give up. If you set yourself on a healthful path that you know is the right one but you eventually reach a point where you feel tired of it, still do not give up. You can do this thing. Pick a habit or two and build it into your day (or evening). Once it is in, start on the next one. The stronger your health foundation, the faster and easier it will be to scale up all the other good habits that will get you where you want to go!

Recommended Reading

- *Why We Sleep: Unlocking the Power of Sleep and Dreams* by Matthew Walker.
- *Salt, Sugar, Fat: How the Food Giants Hooked Us* by Michael Moss.
- *What Makes Olga Run? The Mystery of the 90-Something Track Star and What She Can Teach Us About Living Longer, Happier Lives* by Bruce Grierson.
- *Becoming Odyssa: Adventures on the Appalachian Trail* by Jennifer Pharr Davis.

- *Let Your Mind Run: A Memoir of Thinking My Way to Victory* by Deena Kastor.
- *The Telomere Effect: Living Younger, Healthier, Longer* by Elizabeth Blackburn, PhD and Elissa Epel, PhD.

Strategies for Success: Health

- Get enough sleep and get *good* sleep.
- Take deliberate care of your fuel, hydration, and exercise, to keep your physical health in good form.
- Pay attention to your personal hygiene. In the shower, wash all your skin folds with soap and hot water, and then rinse and dry. Hang up your towel. Put on deodorant.
- Wash and moisturize your face, comb or brush your hair (front and back), brush and floss your teeth, and keep your nails clean and trimmed.
- Take the time to develop your morning routine and evening routine. These bookends to your day set you up for success.
- Book time for your mental health in your calendar and spend your downtime creatively, whether with friends, with a hobby, reading a book, or hiking outdoors. Everyone needs to unplug (screen time does not count).
- Keep up with the spiritual practices you learned growing up. If you did not have a religious or spiritual upbringing, check out the options you have on campus. You might be surprised at the friends you make and the impact this has on your life.

12

Build Your Résumé

You have to make your own luck. You make it in practice and with your
training and conditioning.
—SIMONE BILES

YOUR RÉSUMÉ IS MORE THAN just a list of your work history, education, and awards and honors. You will, of course, list all these things, but your ultimate goal is to show your future employer that you are well rounded and can use your considerable skillset in a variety of capacities.

This chapter is about the areas we have not covered so far that can help you round out your skills and demonstrate a) your ability to work in teams and b) your time management skills.

Note that each of the sections in this chapter outline optional actions which may or may not apply to each year that you are in college for.

Before we begin: I try to ensure every year is accounted for on my résumé. Sometimes a year will be given significant page space (like a graduation year), and sometimes there will just be a short note about it (like in a year where you noted an organization leadership position, or a publication). Regardless of where you go or what you do, ensure every year is accounted for on your résumé so you can show constant growth and learning.

While I always encourage you to cast a wide net while you're in school (so you can determine your strengths and interests), you also do not want to be casting it so wide that it starts to tear and tatter at your commitments. Do not be afraid to push yourself—you are in college, which is a low-stakes environment for good, honest mistakes—but keep watch that you are finishing the items on your growing list of commitments. Your capacity to finish what you start will be the biggest indicator of whether you are overloaded, and will ultimately determine how extensive your résumé is.

Each of the items listed here should help you to develop your soft skills, increase your network, and build friendships with people that you might not otherwise meet. You'll be learning new things with each event that you list on your résumé, and that is why these items should go on it: they serve to develop the whole person.

Clubs and Associations

There are a variety of clubs and associations on every campus and in every discipline. Some clubs are specific to the discipline you are studying (for example, a chemistry club), but there are also clubs representing larger organizations, such as the Society of Petroleum Engineers and the Society of Women Engineers. There are also associations related to the campus, like a Student Association, where you can run for office. You might aspire to be a student leader on campus, and this is a good way to serve in the student Senate, if you will.

There are many more clubs and associations than could be listed here, so look first to your department and then your campus at large. If nothing catches your attention, you can start your own chapter or club. These organizations can be *ad hoc* or highly organized—it's up to you. You just need an idea that will get students involved.

I started a book club years ago, and it was based solely on STEM books.

The book choices had to be nonfiction and at least 50% had to include women either as the subject or as the author. We generally read a book a month. I kept it very structured: we met every week and discussed the assignments I had given out the week prior.

This was one of the most fun clubs I had ever belonged to, and the members made all the difference.

With that said, starting a group takes a tremendous amount of energy and commitment, so if you are new to college, my recommendation is to start with a group or a club that has already been established. Join and get a sense of how the dynamics work. Most universities have a signup day in the main quad or gym for all the clubs, during which you can walk around and find out who does what for which discipline, department, or the university as a whole. I highly recommend attending this event (which usually takes place at the start of the year). Go learn about all the different opportunities that are at your disposal!

You want to be in at least one organization related to your discipline of study. After that, I recommend joining one that is either a campus-wide organization or that has a broader scope than your department club. Being in college is about meeting new people and building your network, after all, and casting your net further will expose you to new opportunities that you simply would not hear about if you only stayed in your department.

There are also student chapters representing national associations that either host or participate in competitions around the country. These competitions are a great deal of fun, and will give you the opportunity to travel. Depending on your university chapter, you may or may not need to fundraise to attend. Several of the groups I've seen compete around the country are quite good, and again, if you compete or place, this is something that will go on your résumé and demonstrate to future employers that you have the ability to (successfully) manage competing time commitments.

Undergraduate Research Projects

Each university runs their undergraduate research program a little bit differently. There are some who have it highly structured, where you submit a project description to the advisor (who determines whether you proceed). Some universities leave it up to the professors in each department to decide whether they want to take on an undergraduate for some of the research. Others support traveling to the national conference for undergraduate researchers, in which case you will have the opportunity to present your research and have your work published in the proceedings.[26]

The reason I recommend trying an undergraduate research project is, it will give you the opportunity to see if you really do like research. In such a project, you will typically work with an advisor or research professor, and you will be able to observe and be mentored in proper research techniques and processes.

You may find that you really enjoy working in a lab or in the field—or you may find that you like none of it! Either way, it is a good idea to participate; doing so will help you narrow down what you realistically want to do after you graduate from college. I, for one, did not know that I would like research as much as I did prior to completing my own undergrad research project, and doing so accordingly altered my planned path forward. I enjoyed the lab work, refining the question, finding an answer that I documented, and presenting my results to a group of my peers.

I will also tell you that everyone who does undergraduate research believes they will continue doing that same line of research when they go on to their master's, and then in their master's when they go on to their PhD. This tells me two things: one, that you really enjoyed what you studied (and that is wonderful); and two, that you need to go find another research project. It is not appropriate to study what you did as an undergraduate, as a master's candidate—and, similarly, as a PhD student, you should not be

---

[26] "Proceedings" is a general title or term that describes all the research presented at a particular conference. They are sometimes peer reviewed, sometimes not. The journal will tell you whether this is the case.

working on the same line of questioning that you did as a master's candidate.

My point here is, if you think you might be interested in research, undergraduate research projects are a wonderful way to proceed *without* committing to something as large as a master's degree.

Try it out and learn as much as you can—and then include it on your résumé!

Attend a Conference

There are many reasons why you should attend professional conferences in your field: meeting potential employers or research collaborators; reconnecting with former colleagues; engaging in round-table conversations about technical advances in your field; getting away from the office; visiting a new city or country; promoting your cutting-edge work; gaining exposure to new and exciting ideas... the list goes on.

Having attended many conferences over the years, my recommendations for getting the most out of your experience are as follows.

Conference Pre-Work

Prior to attending the conference, you should have access to the schedule or program of speakers. Most people (including me) cannot sit through eight solid hours of technical presentations for three straight days, so sketch out a rough agenda for yourself, including:

- Specific talks you are interested in going to, and the corresponding room locations.
- Vendors you want to meet and ask questions to about their services or technology.
- What you want to get out of the conference. A job? Technical updates?

Consider:

- Are you presenting your own paper or abstract and need to locate the room and available technology? Do you need to upload your presentation in the speakers' lounge?
- Is there someone you would like to meet in person? If so, it might be an idea to look ahead and find a coffee shop or quiet area to conduct an informational interview (see Chapter 13).

I also strongly encourage you to stay in the hotel associated with the conference. While this may be slightly more expensive on the front end, the hidden costs of staying in a cheaper hotel further away can be very high. These include:

- The ride fare (plus tip) to and from the hotel and the conference.
- The wasted time spent traveling back and forth (especially if this is a city with which you are unfamiliar).
- The lost opportunities for a) networking and b) bumping into key people in the lobby of the hotel. I cannot tell you how often I have bumped into a speaker or person of interest in the hotel lobby or the elevator!

Plus, your hotel room will be easily accessible during the conference, so should you forget something or spill on your suit, you can easily remedy the situation.

Attending the Conference

Arrive the day before for local conferences, or a day and a half before if you are traveling internationally. Flights get delayed or canceled, and if you worked this hard to get a place at the conference, you do not want to miss it because of something you could have accounted for in advance.

If you are a speaker, plan to keep your routine as close to home as possible. Bring your workout clothes (and do a workout), eat healthful foods, and do not overindulge in alcohol the night before your presentation.

At a minimum, dress business casual. Students, young professionals, and those looking for job opportunities should err on the side of business professional dress. A suit jacket and tie speak professionalism and will never be inappropriate, and those with whom you speak will take notice of your efforts (in a good way). Do not wear hoodies, T-shirts, wrinkled khakis, tennis shoes, white socks with dress shoes, pants/skirts in need of tailoring, or a cap.[27] See Chapter 14 for a full guide on how to dress to impress.

See the "Stupid Questions" section in Chapter 3 for what *not* to ask while in a conference.

In addition to what we've discussed so far, be sure to adhere to this general etiquette:

- If your cellphone rings, stop the ringing immediately and make a discreet exit (or, better yet, have your phone on silent in the first place). Do not start speaking on your cellphone until you are outside the session room.
- For snacks/candy wrappers/cough drops/tissues, quickly rip the wrapper off. There is no way to subtly take off a wrapper, and everyone can hear you doing it, so just get the wrapper off and be done with it.
- Attendees coming and going during sessions is perfectly acceptable. What is not acceptable is when they allow the door to slam upon their entrance or exit. Once you open the door and make your way through, hold it with your hand and slowly bring it closed. The room's attention belongs to the speaker, not you.

Now go and enjoy the conference, my friends! Take good notes, ask questions, hand out your business cards (more on that shortly), and learn a great deal. Your time will be well spent.

---

[27] One conference-accepted oddity, however, is the backpack. Conference attendees usually receive some sort of bag, but many arrive with and use their backpacks (yes, while wearing their business casual dress).

Study Abroad

Depending on your program of study, studying abroad can be a bit tricky. Engineering programs typically have a very ordered class flow that are only offered once a semester, so if you take a semester off to study abroad, you may be set back an entire year. Don't worry, though: there are a couple of workarounds for this problem. First, there are universities that operate on a 4-1-4 schedule. My undergraduate program at Pacific Lutheran University did this, and I studied abroad for two of the J-terms. This worked out well because although I was not in an engineering program, I was an athlete, which meant I could not afford to take a year or a semester off, but taking January off was fine. So, one January, I studied in Norway, and another I spent in India.

As another example, an aerospace engineer I once knew was also minoring in French at the time (see the "Layer On a Minor" section later in this chapter). She decided that she wanted to study abroad for a semester in France to develop her language skills and... well, because living in Paris for a semester would just be fun! It did set her back a year, though, so she took a super senior year and spent five years total in her undergraduate program.[28]

If you look ahead and plan for either your first or second semester in your junior year (and arrange your courses accordingly), you may be able to adjust your schedule such that you can study abroad for a semester (if your university does not offer J-term opportunities). You might also be able to go for a summer.

Do not think that just because most students complete a four-year degree in a specific way, you need to do the same. I promise that extending your studies by one semester or one year will not negatively impact your employability after graduation. In fact, you might improve your career opportunities, since you will have taken the initiative to work your schedule

---

[28] You'll probably know that five years for an engineering undergraduate degree is not uncommon anymore; undergraduate degree programs are getting to the point where you almost need that much time to finish.

a little bit differently, and employers will find this interesting when reviewing their applicants for entry-level positions. The same applies if you are looking to go onto graduate school. As for the military, they will be happy to take you provided you can hit the standards, so you will be fine there, too.

Studying abroad in college is one of the best experiences you can have, and if you can swing it, I highly recommend doing so.

Learn to Code

There are various reasons why you should learn to code, and there are nearly as many codes to learn. Do not fret—you do not need to learn them all—but if you pick one and are proficient at it (or, better yet, stellar), you will set yourself apart from the thousands of other students that will graduating from college the same year as you.

In the following list is a variety of codes, along with what kind of student might find each kind useful. If you are a computer programmer, you can skip this section, as you will already know it, but assuming you're not...

- Python. This is the best for beginners, it runs on just about any system, and it has the most tutorials and online help.
- Visual Basic (VBA). This is great for advancing the uses of Excel, and it can be passed onto non-users for their own use. As an engineer, I wish I had learned VBA early on, as this would have made numerous tedious projects I worked on in Excel much faster, more efficient, and more reliable.
- C and C++. This requires the user to understand how computers work, and is thus best for those studying computer science or something similar.
- LATEX. This is a document code with a hard to beat equation editor. This is best for those looking to write professional-looking documents, research, and scientific papers. I use LATEX before submitting every proposal, because it looks much more professional.

- IDL or MatLab. Unlike the other codes in this list, these ones are proprietary and require a license. Both are meant for heavy data computation. IDL is used by NASA; academia has largely picked up MatLab. If you are interested in numerical modeling, whether you are an earth scientist or a statistician, IDL might be the one for you.

But why would you want to learn how to code?

Well, first, it will help you do your job better. It may also open other opportunities that you never expected, and this is part and parcel of creating a career for yourself that you can both enjoy and excel at. From a high-level perspective, coding will help you reinforce good skills, such as starting out with a plan, identifying any potential trouble areas, debugging or troubleshooting, and reinforcing your attention to detail. Knowing how to code (even if it is not included in your role in a company) will also help you to manage expectations when working with different departments in your organization. It will expedite any number-crunching projects that you have, and will allow you to solve any modeling challenges clients present more creatively.

In short, knowing how to code will help with your cognitive skills and your general abilities, so pick one or two and dedicate yourself to really learning them. Perhaps you could even start a coding club on campus.

The best way to understand how a coding program works is to use it for a real problem you need to solve. I never would have learned LATEX if I had not forced myself to take on a project that arguably could have been done in a Word doc instead. I wanted to understand the ins and outs of this program, and in deciding to do all my writing in LATEX going forward, I not only learned to troubleshoot, but also how to really make each document unique.

There are an abundance of online communities that you can access, and there are resources available anywhere you have the Internet. Plus, you may be surprised by the people you meet on campus who share this same interest.

If you are reading this as an expert, consider helping out a neophyte. You may just create a new convert!

Layer On a Minor

Adding a minor to your major course of study can serve you well. There are a few common choices that pair nicely and stand out to potential employers or graduate advisors.

If you are in one of the pure sciences for your discipline, great work; we need you! I strongly encourage you to consider taking on a math or statistics minor to go with your degree: you will most likely need a graduate degree to go with your bachelor's degree (see the "Graduate School" section in Chapter 15), and all science involves data. Data streams are only going to get larger, and this means going forward, we need scientists who can do numerical modeling and who know how to condition and interpret large swathes of data intelligently and usefully. As a pure scientist transitioning from your undergraduate education to your graduate degree, you will be competing against scientists like yourself who all have the standard suite of required courses, so taking advanced mathematics (either for additional coursework or a full minor) will set you apart substantially from the others in your discipline.

A math or statistics minor will also bear out with scholarships, TA'ships, or research assistantships—meaning this will translate to money in graduate school. Advisors will see that you have gone above and beyond and taken harder courses outside of your discipline. You will not be expected to maintain the same grade point as someone who stayed entirely within their discipline, so do not worry about taking on harder classes that may be (and should be) a challenge for you.

If you are an engineer, you will already have quite a bit of math in your degree program. You may consider taking on additional math courses for a minor. Alternatively, a business minor pairs nicely with engineering, because every project you work on will have an economic component to it.

Should you go on a management track when you are working at your future companies, a business finance or economics minor will help you.

If you are a math major, the world is your oyster! Pick a minor in any science. Demonstrating your ability to apply your spectacular mathematical techniques to a given scientific field will only serve you well.

Recommended Reading

- *Basic Black: The Essential Guide for Getting Ahead at Work (and in Life)* by Cathie Black.
- *American Icon: Alan Mulally and the Fight to Save Ford Motor Company* by Bryce G. Hoffman.
- *The Pursuit of Happyness* by Chris Gardner, with Quincy Troupe.
- *Sam Walton: Made in America* by Sam Walton, with John Huey.

Strategies for Success: Build Your Résumé

- Building out your résumé will not only increase your employability, but will also give you a chance to try new and interesting activities you might not otherwise have considered.
- Join a club or association while in school—one in your department that corresponds to your major, and (if you have the bandwidth) a broader-scope organization, to give you a different view of campus.
- Run for and hold a leadership position in one of your clubs.
- Participate in your campus' undergraduate research program.
- Attend a conference, either as a student or as an author.
- Go abroad for a semester or a J-term. See what's out there!
- Learn to code.
- Take a minor that directly supports your studies and career goals (e.g., math, statistics, business) or take one for fun (e.g., Egyptology, French literature, Scandinavian studies).

13

Preparing for Next Steps

What you do makes a difference, and you have to decide what kind of
difference you want to make.
—JANE GOODALL

O NE OF THE MOST EXCITING aspects of going to college is the
opportunity to imagine the life you can build for yourself and your
family (should you choose to have one). College is when you start
penciling out what you think you want to do three, five, ten, and forty years
down the line.

I say "penciling out" because, of course, the plan is subject to change.
Our lives are full of setbacks and serendipity—we can count on that!—and
we won't always be able to tell one from the other until we've had a chance
to look back at how we responded to it.

As an example: when I lost my job after the oil boom busted, I could only
see it as a setback. All those years of hard work, and suddenly, I was standing
on the curb in the snow, with no idea where to go or what to do. Years later,
however, I can tell you this turned out to be serendipitous: I moved to a
different city, finished two degrees, met and married my husband,
coauthored a textbook, and ran for office. None of these opportunities

would have been possible had I remained in my engineering position.

But for now, let us stay focused on looking ahead. How do you decide where you want to go, what you want to do, and how you want your career to be shaped? Do you go straight into a professional field, or do you go to graduate school?

Did you commit to the military through an ROTC program? Sometimes, officers stay in for their required number of years and then transition to the professional world, and other times, they end up making a lifelong career of the military. The good news is, you'll have much experience from which to draw, and you do not have to make this decision in a vacuum. In fact, I urge you to *not* make this decision in a vacuum. Start assembling a list, notebook, or binder full of ideas.

To assist in this, this chapter discusses the many ways in which you can introduce yourself to people, career options, careers that you'll maybe want to stay away from, subjects you might not have considered studying, and different ways of shaping your career that you could not possibly know from your vantage point as an undergraduate student.

Before we begin, let it be known that I strongly advise you to take every opportunity to introduce yourself to speakers who come in for seminar speeches. I also encourage you to conduct informational interviews with professionals you have located on LinkedIn or during your review of periodicals and papers. We'll talk more about preparing for professional interviews later.

Let's start by looking at ways in which you can articulate your future goals based on the experiences and advice of others.

Seminars and Speaker Series

If you are going through your college program and are enjoying what you are studying, but still aren't totally certain what you want to do after graduation, I encourage you to attend speaker series and seminars. Going to hear those who are practicing in the field can give you ideas. Plus, by attending, you will have the chance to go up and visit that person once the seminar is over.

Universities and departments go to a great deal of trouble to bring in informed, talented speakers, so take full advantage of these opportunities. Attending will only benefit you.

Of course, some talks will be abysmally boring (if not outright painful), but you can learn just as much from a terrible speaker as you can from an excellent one. Pay attention to what they studied, how far they went in their academic background, what they did on their professional track, and what they are presenting on now. When a speaker is being introduced, a biographical sketch is often provided, and this will tell you, in short order, how they got to where they are now. You may well also pick up on interesting research ideas or conversation points.

Occasionally, these speakers will be taken out to lunch or dinner by the hosting department. If you are truly interested in the speaker and their subject matter, I encourage you to see if you can go with them, to extend the conversation. Look ahead to who is speaking on campus and have your résumé or business card ready, printed, and in hand (we'll talk about this in Chapter 14).

I once attended a wonderful lecture by the CEO of a national energy company. I asked the hosting professor if I could go along for dinner, and offered to pay my own way. I wanted to visit with this founder and CEO and ask her how she'd done what she had, and, more importantly, how I could perhaps be involved in her company.

The professor was delighted that a student would want to attend, and said yes.

Here is the interesting part: there were various students who also attended this dinner who also wanted to have a long conversation with this talented geologist and businesswoman. Yet I was the only one who brought my résumé (in a clean, addressed envelope), which I handed to her discreetly after dinner. She later told me that no other students even brought their contact information to exchange with her!

This was a sit-down dinner with a decision maker, so do not miss opportunities like this one. Bring your résumé (see Chapter 12) or your business card (see Chapter 14).

To this day, she and I still keep in touch. While I am not directly involved in her business, she recommended me for a position recently.

Our takeaway here? Always be polite, professional, and prepared. Employers and companies are desperate for those kinds of employees.

Informational Interviews

How do you accelerate your understanding of a job position or company? I first learned about informational interviews (which are highly useful yet rarely used) during my junior year in business school. My first interview was with an international human rights arbitrator and judge. She put herself through law school as a purser on international flights, attending class during the week and handling the on-flight monies between Oregon and somewhere in Russia on weekends. From her I learned that if you want it, you have to get after it.

The purpose of an informational interview is to gather information about a particular job or company from someone who is "in the know". To get you to conduct your interview effectively, let's look at a few key steps next.

Before starting, ask yourself, "What position or company am I interested in? Why?" Next, find someone to interview. Who do you know in that position or company? Be bold! Ask someone you do not know. Interviewing someone you know is safe; interviewing someone you do not know will introduce you to more unknowns (i.e., opportunity!).

Prep Work

- Before scheduling the interview, determine what you are trying to understand (how they got the job; responsibilities; purpose; etc.). If you do not know yet, that is just fine.

- Start writing down a list of questions. Order does not matter.
- Develop open-ended questions.
- Know ahead of time if you would like to do a phone interview or meet for coffee or for a meal (increasing the time commitment).

Asking for the Informational Interview

- Begin by introducing yourself and explaining how you got their name and number, and why you are interested in interviewing them.
- Tell the interviewee approximately how much time you need (fifteen minutes on the phone, twenty to thirty minutes for coffee, or an hour for lunch). Be considerate. If they only have time for coffee, do not push for lunch.

Conducting the Informational Interview

- Dress business casual or business professional.
- Begin by setting the context: why are you interested?
- Ask your prepared questions.
- Have a couple of fun questions ready (e.g., what book is on the interviewee's reading table?).
- Do not be afraid to ask interesting questions. Generally speaking, people love to teach and share their experiences.
- Ask permission to take notes (overwhelmingly, people are flattered and say yes). If permission is granted, *take notes*. I use a dedicated composition notebook for this purpose, and review periodically.
- Offer to pay for the coffee or lunch.

Post-Interview

- Send a handwritten thank-you note within a day (see Chapter 14). Include a comment that resonated with you.
- Keep in touch with the person who shared their time and expertise with you.

Now that you have a more refined idea of your potential company or career path, start looking for your next informational interview. Given what you have learned, why are you interested, and what do you hope to learn from the next one? What do you want to know? Who can share their experiences with you?

Internships and Co-Ops

Internships are a wonderful way to determine if you like a company or industry and to make good money during your summer (see Chapter 7).

You will generally apply and interview for internships during your Fall Career Fair. Occasionally, there are Spring Career Fairs, but those are less common, and it can be more difficult to line up an internship from a Spring Fair. This does not mean you should not try, however; this is how I happened to land my first internship.

Prior to attending a career fair, you will want to have your résumé polished to a solid finish and ready to hand out. Occasionally, companies will not accept résumés, as they believe doing so is a tacit offer of a position. If this is the case, I recommend having business cards on hand to give to the recruiters that come to your career fair (see the "Business Card" section in Chapter 14).

When you go on your internship, I highly recommend keeping a dedicated notebook for the experience. You will want to keep track of everything, from your daily assignments, to the people you meet, to the words of wisdom given to you. I have composition notebooks from each of my internships, and to this day, I go back and reference somebody that I met or a piece of advice that was given to me. These do not need to be fancy notebooks; a simple comp notebook from the bookstore will do.

Remember that every day is an interview, and if you think this is a company you want to work for, then you will want to not only do your best work, but have the best attitude. These are competitive positions, so if you want one, you will need to work for it! Do not be afraid to ask for clarification, guidance, or help, always abide by the safety protocols that are given, and when you have the chance to attend student lunches or lunch-and-learns, ensure you do so. They are not put on for the benefit of the speaker, but for the benefit of the interns.

You will also find that you will make friends with students from around the country, and this is an added bonus. Start building your network now, because these will be the professionals in your cohort.

Professional Interviews

Like every other test, the professional interview is not one to leave to chance. This is the exciting time that comes after you have handed off your résumé and visited recruiters at the job fair. Now is when you will truly get to shine and visit members of the industry in which you might want to work.

Interviewing for a professional position is serious business, but it can (and should) also be a great deal of fun.

Here are my recommendations for a top-notch successful interview.

First, try on your interview outfit before the interview. You want to actually practice a trial-run interview in your outfit for a couple of reasons. Is the waistband so snug that you will be distracted when answering questions (because your circulation is being cut off)? If you are in a skirt suit, does the hemline come up so far as to cause you discomfort? Practice in the shoes and the suit/dress shirt/blouse that you intend to wear, so that you are not surprised at all when your big day comes.

Erring toward conservative dress (rather than daring) is best practice.

When you are running your company, you can wear whatever strikes your mood, but for now, navy, gray, black, and dark brown, with neutral-colored shirts, is the order of the day.

Next, have a friend interview you. Answer questions that you can easily find on the Internet. Interview questions are not generally intended to be "gotcha" questions, and you can find lists of them on many websites.

Remember that the point is not to have a precise answer for every question, but to be comfortable with answering an odd assortment of questions as they are asked of you. You know you will be asked about your application and résumé, but you may also be asked how you would behave in certain settings. Many times, there are not any "right" answers, but there can be wrong answers (any answer that indicates you are a misanthrope will be a red flag for the recruiter!).

You want to put on your best self for the interview, and this means going through some of the motions ahead of time, even if you feel a little silly doing it.

I have worked with students who seriously struggle with interviews, and more often than not, this comes down to confidence. I understand how hard this can be for some, and I want you to know that the more you practice, the easier it becomes—truly. Always have a good clean joke or anecdote that you can relay about yourself. Employers like to know that you have a good sense of humor. Work hours can be long, and we want to know that we can get along with our colleagues under any circumstance.

When you are interviewing, sit up straight and keep your hands folded lightly in your lap—or, if you are holding your portfolio with copies of your résumé in it, that is fine, too. Remember to smile periodically during your interview and shake hands firmly at the end. Look your interviewers in the eye and thank them formally by their title and then name (Dr., Ms., or Mr.). When you are finished, if you have their business cards or contact details, certainly send a quick thank-you to their email address—or, if you are able to (and better yet), send a handwritten thank-you note to them at their office (see the "Thank-You Cards" section in Chapter 14).

Job versus Career

When you're in school, you will hear a lot of professors—well-intentioned professors—talk about the importance of getting a job. I contend (on the premise that any paid gig is a job) that if you are going to go to all the trouble, time, and effort of obtaining a four-year bachelor's degree in a STEM field and are not headed to graduate school or the military, then what you really should be aiming for is a *career* postgraduation.

You might have had jobs in high school after school or during the summer (certainly, you should during the summers in college), and these jobs were likely good ones: they either taught or reinforced important habits like working with others, timeliness, finishing what you start, and organizational skills. All these habits are transferable to your future career. What these jobs did not offer you, however, was the opportunity for upward mobility.

I would loosely characterize a "career" as a role in which you have more influence and control over the direction of your trajectory, and in which your trajectory is upward. You do not think about a job after you leave work, whereas a career tends to go home with you. A job is generally paid hourly, whereas a career is generally, but not always, paid with a salary.

Some larger companies will allow you to choose between a technical track, a project management track, or a business management track.

- The technical track is a wonderful opportunity for those who are more inclined toward research and development, and who want to stay on the technical side. This track generally develops from positions following your entry-level hire, and allows technically minded professionals to still achieve advancement and promotion in responsibilities, title, and pay.
- The project management track is mostly available in really large engineering firms. Project management is a vital role; someone has to run these projects and keep everyone on time and on budget! If you are detail oriented, can juggle competing interests without losing your cool, and can get team members to "eat their peas and broccoli" (as a friend of mine likes to say), this track may be for you. The ability to manage

large projects is a valuable skill, and the rewards can be tremendous. Project managers can continue managing projects for a career, or they might jump over to the business management track after a while.

- Business management tracks are also born out of positions following the entry-level phase. These roles train you for management, a vice presidency, or even the C-Suite. If you are on this track, you will likely rotate through various aspects of the company after you complete your entry-level position, and you will likely need to complete additional training in finance (think MBA) and to take on non-STEM-related roles (human resources or government relations, for example) while you learn about the various aspects of your company.

All three tracks lead to an exciting career, so take the long view.

Does this mean forgoing a job completely? Not at all. A job is very important for paying the bills right now. No matter what job you have while you are in school, note it on your résumé. Your summer or during-school jobs do not need to be related to your field of study; employers will just be excited to see that you are working and that someone impartial can attest to your work habits.

As should go without saying, you need to treat your supervisor respectfully so that they can provide a recommendation for you when you apply for your next career move. Remember, every day is an interview. Just because you are not in the position you aspire to be in right now does not mean you will not be at some point, and you will need people to help you get there. Do good work in your job for now, and keep your eye on career opportunities after you graduate.

Recommended Reading

- *Business Etiquette for Students and New Professionals* by Mary Crane.
- *How to Work a Room: The Ultimate Guide to Making Lasting*

Connections—In Person and Online by Susan RoAne.
- *How Will You Measure Your Life?* by Clayton Christiansen.

Strategies for Success: Preparing for Next Steps

- Take inventory of the relationships you built over your college career and note those whom you would like to interview for information, to guide your career decisions.
- Prepare diligently prior to your informational interview. Have your questions ready!
- While on your internship or co-op, remember that every day is an interview both for you and for the company.
- If you're looking to build a career, first consider, do you like the work that comes with that path? The corporate culture? The colleagues?
- Consider what track (technical, project management, or business management) you might be interested in, and interview professionals in those areas.
- Send thank-you notes after any and all interviews.

14

Professionalism

There are three things to remember about being a Starship Captain: keep your shirt tucked in, go down with the ship, and never abandon a member of your crew.

—CAPTAIN KATHRYN JANEWAY

PROFESSIONALISM IS NOT JUST ABOUT how you dress, although it does include it. Professionalism is about how you conduct yourself in a variety of circumstances.

In the quote at the beginning of this chapter, Captain Janeway makes three points to remember: keeping your shirt tucked in (i.e., keeping it together even when those around you are not), going down with your ship (i.e., staying on board your mission or task), and never abandoning a member of your crew (which reminds us all that our colleagues matter and we should look out for one another).

Do you have a calm, levelheaded approach when everyone around you is flapping their wings and feathers are flying? Then you are a professional. After all, crises and emergencies happen. Frustrating events (like getting a flat tire on your way to work) happen. Coffee will sometimes be accidentally dumped down the front of your shirt. Not losing your cool and keeping your focus is the best way to demonstrate that you are a professional. Do not allow your appearance to slide. Keep your shirt tucked in.

Men and women cry at work. I have seen this happen, and have been on the receiving end of it. If such an occasion arises, discreetly go to your office or the bathroom and attend to what is needed. Then, wipe your nose and get back out there. Every person I know has been dealt a blow at work professionally, just like every person I knew by the time I hit adulthood had had to deal with trying personal circumstances.

People are generally sympathetic, and this humanity is one of our best attributes, but you also want to be that person who continues moving forward even when there are reasons to stop in the middle. If a situation as serious as the death of a loved one arises, of course you need to take time off, but these events are few and far between. You do not want to be the person making a mountain out of a molehill, so before flapping around with a lot of drama and feathers, ascertain whether this is the appropriate response.

Professionalism is also about good manners, clean language, and treating those around you with respect. Professionalism is about a can-do attitude and being able to recognize when you are starting to overwork (and taking the time off that you need accordingly). Professionalism is not synonymous with being addicted to work, sleep deprived, or burned out.

Keep your workspace tidy, and do not overdo the personal mementos of home and clutter. This goes for your car, too: don't leave the passenger footwell full of empty paper coffee cups, sandwich wrappers, or other bits of life detritus. You may be called on to take the VP to lunch, and this will be your moment to shine with your brilliant intelligence and witty conversation. The very last thing you want is for the VP to have to nudge aside your garbage so they have room for their feet.

Let's look at a few other components to being a professional so that when you go forward, you always have your shirt tucked in—literally and metaphorically.

Dress

Unless you are the president of the university or one of the faculty, I am not suggesting that you wear two-piece business professional attire every day in college. That would be silly, and would defeat some of the fun of college. Do, however, wear clean clothes and look somewhat put together every day. Have your shirt tucked in, a belt in your trousers, and nothing obscene on your clothing.

Do not wear soiled workout clothes to class. No one should have to smell that. And yes, it is obvious that you pulled a T-shirt off the floor (or out of the hamper) and put it on, so do not do that. To step up your dressing game just a notch, gentlemen, wear a polo, and ladies, a sweater. You can pair with jeans or casual trousers, and you will still be a step ahead of the jeans and T-shirt crowd. When professors are looking to put one of their students forward, you want to be that student, so look the part.

College is your training ground for your next steps in life, so you want to make sure that you are putting good dress habits into practice. If you wake up feeling a bit blue or mentally under the weather, put on a brighter color rather than gray or black clothing. Dress a little up rather than a little down. You will notice that your mood lifts, too.

Business Professional

This section on dress is aimed specifically at helping you for a professional interview in a STEM career field. For readers who see themselves as fashion aficionados, feel free to skip to the next section.

For the rest of us, when interviewing for a professional position after graduation (whether a summer internship or an entry-level position), there is a business professional uniform of sorts that is expected. If you are not particularly creative, I have good news for you: the colors will all be very neutral. If you *are* a deeply creative sort, this will be a bit of a restriction, but I promise it will help get you to the end goal: a wonderful paying position in your new career.

For both men and women, your color scheme should be black, dark gray, navy, and earth tones. This will go for your blazer, suit, or trousers. The easiest suit to put together is one that matches the jacket and trousers or jacket and skirt. These can be expensive and are not expected of an undergraduate, so if this is beyond your means, do not worry; it is not a strike against you. One of the advantages now with the Internet is the availability of many used clothing sites, and I still avail myself of these wonderful finds.

Rather than just using the sizes given, take the extra time to measure yourself (bust/chest, natural waist, low waist, hips, and inseam are typical measurements), and then confirm which garment size corresponds to your measurements. Sizing in America can be unreliable, so this is worth the effort. Note that sometimes, the measurements given for a piece of clothing are the measurements for the person it should fit *or* the measurements of the garment itself.

Take the measurements of the pieces that are already in your closet that fit well (dress shirt or blouse, slacks, or skirt). This will give you a solid starting point for your new/used purchase.

Gentlemen, if you wear a nice navy or gray blazer with clean khakis or earth-colored trousers, a white button-down long-sleeved shirt, a tie, dark dress socks that stay midway up your shin (not ankle socks or white tube socks), and polished leather shoes, you are good to go. It really is that straightforward.

Ladies, if you want to wear a professional dress (these are generally made of a heavier material, like wool or heavy cotton), I still recommend putting a blazer over the top. You can also do the same colors I mentioned before: a gray or navy blazer with an earth-tone skirt and a white button-down blouse. Wear lower heels (no more than two inches high) and hose/tights. The one exception to the hose rule is in the southern states, if you are interviewing in August when the temperature out is sweltering. If that's the case, feel free to skip them.

One question I often receive is, "Do I need to have a different interview suit or outfit for each day of the week during career fair week?" and this is because sometimes, you will have two, three, or even four interviews on different days. The answer to this question is no, you do not. The only priority is that you are clean and pressed. This is not something that you want to go into debt for, if you can help it! Every recruiter understands what it means to be an undergraduate, and if you present yourself as clean and tidy and, better yet, frugal, the recruiters will be that much more impressed.

Gentlemen, do not wear your dress slacks from your eighth-grade graduation with the white pockets popping out on the sides, even if you can squeeze into them. Those pants are too small, and need to be passed onto the next owner. Ladies, you need at least an inch of movement in your skirts or trousers, and no gaps on the bottom portion of your blouse. Small studs for earrings are strongly encouraged, as is neatly brushed (if you have straight hair) or pulled back hair. Don't forget the back of your head as well. Check front *and* back with the mirror. If you have curly hair, do not feel compelled to straighten it. Look at others in your field and further up the career ladder to see the range of possibilities for hair styling, and then choose a style that works for you. The key is to keep it out of your face while you are interviewing and to minimize distractions. Minimal makeup is fine, but remember that you are there to show off your intelligence and problem solving acumen, not to audition for a reality TV show.

As mentioned in the section on professional interviews in Chapter 13, try your outfit on ahead of time and make certain it works and is hemmed properly. If your pant legs are too long, take them to a dry cleaner who does tailoring and have them taken up. It should cost you between eight and twelve dollars to have this done, and it is worth every penny. If you need the waist taken in, then do so; otherwise, make certain you wear a belt so that you are not constantly hitching up your trousers. Practice sitting, standing, and presenting in your interview suit. Practice walking in high heels if you have not done so before.

Whether you are a man or a woman, my recommendation before purchasing dress shoes is to try them on at the end of the day so that your feet are swollen from the day's events. Make certain they do not pinch and are not too small. If your shoes fit then, you will not have any problems.

It goes without saying that you should have washed your hands, trimmed your nails, and dug any dirt or grease from lab experiments out from under your cuticle beds. Brush and floss your teeth before you interview so that your interviewers are not surprised by a misplaced raisin masquerading as a lost tooth.

When you are dressed up, clean, and pressed, you perform your best. You look great. Go crush that interview.

Communication

Communicating with your professors, teaching assistants, university staff, or administrators always requires professional comportment. Do not be overly familiar with those who may one day be writing a letter of recommendation for you, or that you would want to put your name forward to for some honor or accolade. As mentioned in Chapter 3's "Course Behavior" section, do not use foul or coarse language, do not make offensive or inappropriate jokes, treat those around you with respect, and address those who have a title with their title.

Unless they specifically ask you to use their first name, always err on the side of deference. If you have a professor who earned their doctorate, address them as "Dr.", not as "Mr.", "Mrs.", or "Ms.". Also address the president or provost as such, even though they probably also have earned their doctorates. When addressing a vice president, those titles tend to be cumbersome, so it is appropriate to use "Dr." for brevity's sake. You can use the title "Dean" or "Dr." depending on their preference, and for department heads or department chairs, "Dr." is also fine.

Institutions vary in their level of formality, so pay attention to what others around you are doing. If you are asked to address someone in a specific manner, then do so.

Email

Full disclosure: I loathe email. The minute I can afford to hire someone to handle all my emails, I will do so, and I will pay them gladly. Until then, here are the best practices for email.

First, try to respond within twenty-four hours of receiving an email, even if it is just to say, "Received and I'm working on it." Let the person know that their email did not disappear into oblivion, never to be seen again.

Second, when addressing professionals and those who outrank you, always include a salutation and their name. You do not want to present yourself as someone who is unduly or inappropriately familiar. Never begin an email with, "Hey."

Third, keep it brief. Everyone has inboxes full of emails, and part of what slows down the response time is having to slog through interminable epistles. If it is faster to have a phone conversation, then send a quick reply telling them such and find a time for one, or just pick up the phone and call them. If someone must scroll more than once to read the entirety of your email, it is far too long. Feel free to use bullet points.

Next, do not fire off an email when you are in an agitated state. Once it is out there, it is out there. If you are emotional about the matter at hand, draft out the email in a Word doc, where there is no hope of you accidentally sending it. You can always copy-and-paste into the body of an email later if, once you have calmed down, you still believe the email to be appropriate.

Last, if you are emailing a professor, recruiter, or boss, use professional language. I cannot overstate the importance of this. I have seen emails sent to professors that sound like they came out of a B-grade reality TV show. Comedic effect aside, it is disrespectful, and will not gain you favor or grace with your professor. They will certainly not be inclined to write letters of recommendation on your behalf, or to put your name forward for scholarships, opportunities, or awards.

Thank-You Cards

This is such a great idea that I wish it was mine! Years ago, when attending a scholarship awards ceremony and training, we had a speaker who talked about a variety of etiquette rules for professionals. One of the standbys was writing thank-you cards. This man's idea for thank-you cards went as follows: go to the bookstore and buy a pack or two of cards that have your school logo on them. Not only are they professional in quality and appearance, but you will also be reinforcing your connection to your esteemed university by using them. If you are in an engineering program that is renowned at your university, for example, and you send a thank-you card to an informational interviewer or professional interviewer with your school's logo on it, you are reinforcing your abilities and competence based on your school's reputation.

Thank-you cards are not a new or unheard-of thing, but the part I absolutely loved (and I wish I had his name so I could thank him for this great idea) is the fact that while in college, you are in the process of branding yourself. You are presumably proud of where you are studying, and want to be associated with it. Universities go to a lot of work to develop strong Alumni Associations and networks, so in using your school's logo, you are tying yourself to that institution.

Be proud of where you go to school and use this simple gesture to keep up with all the other wonderful alumni who have graduated from your institution. It is a network, so make great use of it and keep yourself included in it.

Business Cards

You can make your own business cards for very little cost by using pre-perforated sheets and designing and printing your own.

I recommend keeping them very simple. On the front, you will want to include items like your name, your mobile number, and your email address.

You may also want to include your LinkedIn profile handle, if that is up to speed, as well as a personal website or blog, if you have one.

Remember, everything you put on the Internet is fair game, and companies often spot check via a quick search on your name. Professionals have lost positions over their social media feeds, so be discerning and diligent about what you put out there.

If you are in graduate school, you *need* business cards. As mentioned previously, recruiters from large companies can no longer accept résumés at career fairs because doing so could convey a tacit job offer, and business cards provide tidy a solution to this problem: if you leave the recruiter with your name, contact information, and (if you are savvy) half a dozen key résumé points, then the recruiter did not break HR rules, and you made a professional impression. This means you are in business.

Recommended Business Card Details

- School name and logo. School logos are typically trademarked, so you may need to get permission or use a print shop which has the official logo. This also saves you the consternation of developing your own logo.
- Your name (in a slightly larger font than everything else)
- Your title (e.g., master's candidate, PhD candidate, etc.)
- Department (and then I use the department's address, not my personal address)
- Mobile number
- Email address
- LinkedIn profile handle (if you have one)
- Website (if you have one)

On the back of your business card, you can include your top three to six skills, with bullet points for emphasis.

When, Where, How, and How Many

- When: For undergraduates, as soon as you would like to get them done. For graduate students in particular, as soon as you pass your exams and move from student to candidate, print your cards. I had mine designed and ready to print so that the day I passed my exams, I could call the print shop and place the order. Do this, and employers and recruiters will know you are not just treading water; they'll know you are serious about your academic and professional ambitions.

- Where: Hand your cards out at every conference, include one in every thank-you card, and give one to every speaker who is even remotely related to your work. Get your cards (and your name and contact information) out there! At a recent conference, I wrote, "Thank you for translating!" at the top of a business card and handed it the woman who interceded for me in Russian. She returned her business card with a response at the top, and we have kept in touch since. Hand your cards out everywhere and include them in professional correspondence.

- How: When handing off your card, make a verbal note about one of the details, whether your cell number, school email, or the spelling of your name. Flip the card over and highlight one of the points on the back. Get the receiver of your card to stop for a moment and look at your card while you are introducing yourself or your idea. This will feel awkward at first, but do not shortchange yourself. Studies show conclusively that when someone both sees and hears about an idea (or person or product), that idea stays with them for much longer and much more fully than if they just heard about it *or* just saw it. Look the person in the eye, smile, shake their hand, and hand them your card while pointing some detail on the card out.

- How Many: Print a minimum of five hundred cards. Any fewer than that, and you will be stingy when handing them out. The point of marketing yourself is to not be discreet, reserved, timid, or parsimonious. If you have an abundance of cards, you will not hoard them. It only takes one card to get you the connection you need to get one job offer. The problem is, you do not know which card will garner you the job offer(s), so hand them all out. For the price of five hundred business cards, two competing job offers are beyond compare. You need to constantly be selling.

Attitude

In this chapter, we've covered how to present yourself (whether in a professional interview setting or in a day-to-day setting), your manners and language, keeping your workspace tidy, and various forms of communication. This section (on attitude) describes how your overall person should be when you are interacting with people professionally.

You can do all the steps we've covered in this chapter so far correctly, but if you have a constantly sour disposition, then you will not be the go-to person for new projects, tasks, or opportunities. When supervisors or professors are looking for someone to represent the company, department, or university, they will look for somebody who meets all the aforementioned criteria *and* has a can-do attitude.

In the beginning of your career, you will probably do more than is required or necessary. This is the right thing to do. As you move further on in your career, you'll be able to trim back and negotiate your work output for your compensation. At the start of your journey, though, it is normal to accept greater challenges even without greater compensation, as these challenges will give you considerable opportunity for professional growth— and this professional growth will leapfrog you past colleagues and peers who were not given the same opportunities.

The only way to really make yourself shine is to consistently put yourself in front of superiors or supervisors who make the decisions, and to make it clear that you are open to taking on challenges. Even better if you take on a project or challenge that no one else wants to do and you knock it out of the park! Opportunities like this are rare, but when they come along, do not be afraid to take them on and really showcase your abilities.

Your attitude is something that you control every single day. Take ownership over your day, work product, and attitude toward your responsibilities. This behavior will set you apart tremendously from the bulk of people who slog through their day. Your peers and superiors alike will wonder why you have such a great attitude, and the reality will simply be that you *chose* to have one.

Go get 'em!

Recommended Reading

- *Executive Presence: The Missing Link Between Merit and Success* by Sylvia Ann Hewlett.
- *My American Journey: An Autobiography* by Colin L. Powell.

Strategies for Success: Professionalism

- Appearance matters. Dress a little up rather than a little down.
- Have a professional outfit that fits and is clean, pressed, and ready to go.
- Communicate professionally via email as well as in person. Do not use overly familiar names or terms unless you are explicitly invited to do so.
- Send thank-you notes regularly. Buy university-branded cards in your bookstore to reinforce your affiliation with your great institution.
- Make business cards and hand them out liberally.
- A good attitude goes a long way and covers a multitude of sins. Keep a positive attitude in every circumstance.

15

After Graduation

If you're not making some notable mistakes along the way, you're certainly not taking enough business and career chances.
—SALLIE KRAWCHECK

THE TIME TO CONSIDER WHAT you will be doing after graduation is *not* after graduation, nor is it the month or two leading up to graduation. You want to be considering your options for postgraduation at least by your junior year of college. I make this recommendation because I did not do this, at least not past my Fulbright Scholarship in Norway, and I regretted it.

When I returned from my year in Norway, I did not have a clear path in front of me; all I knew was that I wanted to work in energy. I worked for the summer as an ice cream taste tester (yes, really) and then moved out to the Pacific Northwest (PNW) to work as a legal secretary. At the time, I thought—no, genuinely believed—that I was going to be an attorney in the energy industry.

I attended Duke for a master's in theological studies and opted to let law school go for the time being, and to go to work instead. Back to the PNW I went.

This time period was the tech bust of 2003, when there were no jobs, never mind careers. So, I started working as a contract employee receptionist at a major utility company. From there, I worked my way to a sales position in a lighting agency selling whole home lighting controls and, finally, to a position as a specialty pharmaceutical representative for a mid-sized company.

After that, I went back to school to study engineering. That part of my career, at least, was planned.

My hope is that planning (or at least sketching out) possible routes for forward movement will make things a bit smoother for you. Starting this process in your junior year may sound excessive, but there are a couple of good reasons for this recommendation. The first (and best) reason is the fact that good planning takes time. Sometimes (as you just read from my backstory!), it takes us a while to figure out where we are supposed to be and what it is exactly that we are supposed to be doing. If you do not get it all right straight out of the gate, you will always have a seat next to me, but it's still nice to minimize surprises where possible!

If you are undecided about where you want to go after graduation, this is when you need to start doing your research. If you are trying to decide between graduate school and going to work for a few years first, this is the time to start weighing up your options, particularly about who will be paying for your advanced degrees. You may have put yourself through school on an ROTC scholarship, in which case your commitment will have you spending the first few years out of school in the military. If you are interested (and depending on their needs), the military may pay for you to go to graduate school, too.

Moreover, there are prestigious scholarships which may or may not involve graduate school, along with a few other options, like volunteering with an international organization.

Many of these adventures require an application and a fair amount of lead time. This is why it is important that you start looking at what you think you might want to do and where you might want to go ahead of time.

When you come out of your undergraduate degree, you will enter a fantastic time with a tremendous amount of freedom, so swing for the fences.[29] If you think you might want to live somewhere new or go do something completely different for a while, this is absolutely the time to do it. If you want to go teach English in a foreign country, do it now. If you want to travel and spend the remainder of your hard-earned internship money seeing the world, do it now. If you want to spend a year in service and have some of your student loan debt reduced, get ready, go!

Let's look in a little more in depth at your postgraduation opportunities.

Professional Life

When I talk about "professional life", I'm talking about any full-time position you have been hired for. This could be in the private sector, or with a local, state, or federal agency—or even perhaps a non-profit agency.

If your plan is to go to work after you graduate, I strongly recommend that you at least obtain an internship during school, or participate in extracurricular activities—or, preferably, both. After visiting the career services offices on various campuses, I can tell you that the students who have the hardest time finding placement (and, more likely, end up not being able to find any placement in their field of choice at all) are the students who did not obtain an internship or participate in extracurricular activities.

In the previous chapter (in which we covered building your résumé), we discussed that the reason why you should build your résumé is so that employers know that you are capable of interacting with other people, have soft skills, can manage your time with competing interests, and have multiple dimensions. Do not fall into the trap of thinking that the only thing to do in college is study hard for your classes. That is important, but if that is *all* you do, then you will have an extremely difficult time finding career opportunities after graduation.[30]

When applying for positions, again, cast a wide net. You may find the

---

[29] I've included a few adventure books at the end of this chapter to whet your appetite!
[30] See Chapters 12 and 13 for further information on the professional life.

position description interesting, only to discover that the actual work or corporate culture is not what you thought it would be. On the other hand, you may take a position that initially does not look like a good fit, only to realize that you really enjoy the work. Companies will do research on you before they make an offer, and you should do the same! Read articles about their work. Read their mission statement and the letter to investors. Conduct an informational interview with someone who has worked there for a few years. Get to know the companies you think are interesting, and then go interview.

Graduate School

Should You Go to Graduate School?

Whether or not you should go to graduate school depends on a couple of factors. First, do you even want to go to graduate school? If you think you might but now is perhaps not the right time, then working for a few years to save money and pay some of your debt is a solid idea. This period may also help you determine what exactly you want to study in graduate school.

You may or may not find a company that is willing to pay for your graduate work. If you do, go for it, though I also caution you that working and attending graduate school full-time, while doable, is incredibly difficult.

Something else that can determine whether people go back to school for an advanced degree is the economy. When the economy slows down or there are layoffs due to low mineral or oil prices, geoscience departments see an influx in graduate school applications. This is neither good nor bad; it is just how people choose to differentiate themselves so they are ready when the price of minerals or petroleum comes back up again.

You may also decide you want to change career directions while you're in graduate school, and that is just fine, too. You do not need to pick the same graduate discipline as your bachelor's degree (e.g., mechanical engineering for a bachelor's degree and then a more refined focus into fluid mechanics for a master's degree). If you are too far afield, though, you may need to fulfil a few prerequisites before you are formally admitted to the graduate program (e.g., a bachelor's degree in business to a master's degree in petroleum engineering[31]).

There is one important thing to note about studying the pure sciences and pure mathematics for an undergraduate degree: while you can find employment after you graduate, you should plan on obtaining a master's degree shortly after your undergraduate degree if you want to find a career opportunity as opposed to a tech job, whether in physics, chemistry, geology, biology, or mathematics. There are times when labs need competent techs, particularly if the mining or oil fields are booming, and in this case, you will be able to work as a field tech, but these are likely not the career opportunities you envisioned when you first started your degree. Engineering programs and business programs, on the other hand, are designed for you to be able to obtain a career after you finish your bachelor's degree, so you do not need to plan on earning an advanced degree for either of those (unless you are specifically interested in doing so).

Call me biased, but I think mixing and matching business and engineering is a dynamite combination! Business and science is also a great blend, but we do not hear about this one as much. Mathematics, meanwhile, underpins everything—statistics; economics; finance; science; engineering—so mathematicians will always have ample choice for graduate programs!

Picking a Graduate School

So, how to pick a graduate school?

Common advice for ideal school and program selection goes something like this: survey a number of graduate schools and focus on the brand name schools; send letters of introduction; pay an in-person visit, meeting with

---

[31] This is what I did, and I needed to make up for quite a few deficiencies.

the professors to determine if your personalities and research topics match; and send out a unique application for each and every graduate school.

If you have the time and resources for this approach, go for it!

Here is how I picked my STEM graduate schools: I filtered by location, reputation for subject of interest, and financing options. There are fine institutions in every nook and cranny of this planet, so I had to filter down a little, and I knew I wanted to be back in the Rocky Mountain Front. I knew Montana Tech and South Dakota Mines had fantastic reputations for their respective programs, I knew I wanted to live and work in that area of the country, and I knew their programs were (and still are) very good value. Even better, I forged tremendous professional relationships, made lifelong friends, and continue to be involved postgraduation.

How you select your graduate program is up to you. There are many variables: quality of the institution; location; size; advisor availability; personal or family considerations. For example, you may have a spouse who has specific career needs, or children with educational requirements.

During the selection process, make sure you consider how you will fund your graduate education. There is no one right answer, so take some time to map out what matters to you, and then go after your applications. Just make sure to get them in before the deadline so you do not miss out on scholarships and teaching assistantships!

Military

Typically, once you have earned your bachelor's degree, you enter the military as an officer. There are a couple of other unique occasions where you would enlist, but I will leave that for a conversation between you and your recruiter.

Note that you do not have to have gone through ROTC to be commissioned in one of the branches of the military. Depending on your specialty, all branches of the military are interested in STEM fields, so you would likely have your pick, if this path is for you.

In the U.S., the branches that generally support graduate degrees are the Air Force, Navy, and Army. The Marines and Coast Guard do, too, but to a lesser extent. Decide on the capacity in which you want to serve and visit recruiters to determine if this is the path you want to take for the next few years—or possibly as a career.

There are many exciting opportunities in the military for engineers, scientists, mathematicians, and technologists. You may want to see the world and subsequently come out with a very solid international work history that is readily transferable to the professional world. Depending on their needs, the military might also provide signing bonuses and cover the cost of tuition, so ask about those options when you are visiting the recruiter.

If you think this option may be of interest, you will want to pay attention to the physical standards required by the branch you will be serving in. While the standards remain high, the goal of the military is to recruit the best and brightest, so if you need help getting your pushups, sit-ups, and runs in, they will help you to achieve your goals.

If you expect to obtain a security clearance, you will need to be (and remain free) of illicit substances.

A military/veteran background can also set you up to have ample choice for posts in other governmental agencies post-military: Department of State; Department of Defense as a civilian; many of the national labs; NASA, or adjacent agencies... these are just a few that come to mind.

Other Paths

When visiting the career services department, you will find that after the first big three (professional life, graduate school, and the military), there are many other options open to you after you graduate. Far fewer students avail themselves of these options than the other three, but you might find it interesting to learn about them as you research your future.

The first of these options is prestigious scholarships. I have provided a list of these in the next subsection. Some involve going to graduate school, while others are a standalone year or two to study in a foreign country or at a foreign institution.

Volunteer programs are a wonderful way to see the world and the United States *and* to give back to your country, or help those who are less fortunate around the globe.

There are also public service loan forgiveness programs, which will help you pay your debt in exchange for time served in an economically depressed community.

There are many options available to you, and these are just a few of them.

Prestigious Scholarships

Start looking at these early—as in, more than a year in advance. You will need to plan strategically to win one. And yes, this is doable. Here are a few to look at, depending on your specific interests:

- Rhodes Scholarship.
- Fulbright Scholarship.
- Marshall Scholarship.
- Gates Cambridge Scholarship.
- George Mitchell Scholarship.
- Winston Churchill Scholarship.
- Harry S. Truman Scholarship.

Volunteer Programs

In the U.S., there are programs like Teach For America, AmeriCorps, and Peace Corps. These organizations give you a chance to volunteer (and work with other volunteers!), go somewhere new, and work on something quite different from anything you will have done in your life so far. At the end of your commitment, you will have a solid segment on your résumé speaking into your credibility, reliability, and experience, and, even better, you will bring back more than a few really good stories!

Public Service Loan Forgiveness

These programs are available for a variety of 501(c)3 nonprofit organizations and some government entities. You will need to have made one hundred and twenty payments, and your loans cannot be in default if they are to be forgiven.

Considering the rules are constantly in flux, I recommend looking at your options online before taking this route. In any event, you will need to have made loan payments for ten years before they can be forgiven, so this will not be an option immediately after college, but one that you can make a note of for later down the road.

In general, working in lower income communities (particularly as an educator) will allow you to avail yourself of this option. Our communities need good science and math teachers, so if that is of interest to you, look up public service loan forgiveness programs online and start putting a plan together!

Cannot Decide?

What if you have analysis paralysis and cannot decide, or just need to catch your breath?

If analysis paralysis is the problem, make your lists. Pick a couple of trusted advisors, parents, or mentors, and go through the pros and cons as they relate to your priorities and values. It is doubtful that you will make a wrong decision, and if you do, so what? Work hard, learn what you can, and try again.

If you just need to catch your breath, take a good, hard look at your finances. That will determine how imminently you need to earn a paycheck.

On occasion, I meet people who pushed hard in high school and college, and so had nothing left in the tank for the next round. Is it okay to press pause, go take a job that pays just enough to cover the bills, and figure out what you want to do with the rest of your life while on the job? Generally speaking, yes. Set a deadline for yourself—six months or a year out—so you do not grow complacent, and then put it out of your mind. Take some time to catch your breath. When your deadline arrives, reevaluate. Do not go to graduate school if you are in the midst of feeling burned out from school. You will not like it, and there is a high probability that you will not finish. Do not take on the debt or time if you are not certain you can finish.

I also recommend not having any gaps on your résumé exceeding one month. You don't want to have to try and explain to your hiring recruiter that you spent six months couch surfing, with nothing to show for it except an astounding understanding of [insert binge-watched show storylines here]! Keep in touch with your campus career development office so they can help you with new opportunities.

If it does not resonate deeply with you, doing what others think you should do (whether it is one of the big three we have discussed or another path) is likely not the right way forward. You will make mistakes, hit a few bumps in the road, and even make a U-turn if you're lucky. Take advice given to you by those who love you and have your best interests at heart, and then decide for yourself. We can all manage someone else's life much better than our own. I get it. And sometimes, those giving advice can be quite... insistent. Thank them for their expertise and interest in your future. Then, square your shoulders, lift your chin, and sashay/march/sail forward into the great unknown that is *your* future. You've got this.

Recommended Reading

- *Call Sign Chaos: Learning to Lead* by Jim Mattis, with Bing West.
- *Delete the Adjective: A Soldier's Adventures in Ranger School* by Lisa Jaster.

- *Getting What You Came For: The Smart Student's Guide to Earning a Master's or Ph.D.* by Robert L. Peters.
- *Kon-Tiki: Across the Pacific by Raft* by Thor Heyerdahl.
- *The Last Place on Earth: Scott and Amundsen's Race to the South Pole* by Roland Huntford.
- *The Magic of Thinking Big* by David J. Schwartz.
- *West With The Night* by Beryl Markham.

Strategies for Success: Next Steps

- Start planning for postgraduation during your junior year of college.
- A professional career is one option. Start building your network and land an internship.
- Graduate school is another option. Participate in an undergraduate research program, attend a conference, or submit an abstract or poster to present.
- If you are ROTC or looking to join the military, speak to the campus commandant or recruiter and follow their recommendations.
- You can also apply for prestigious scholarships, multiyear volunteer programs, or public service, with an end goal of student loan forgiveness.
- The world is waiting for you. Get out there and get after it!

16

Business Plan for Life

If you don't like the road you're walking, start paving another one.
—DOLLY PARTON

AS YOU GROW IN YOUR abilities and responsibilities, you will want to take on new and exciting challenges—and, good news for us, there are always new opportunities to consider. But with limited time and funding, how do you decide?

If you have participated in any of the business programs and competitions on your campus, or if you have taken a business class or two, you might have come across a business plan. You have probably already created something similar to a business plan, even if you did not take business courses. In most STEM degree programs, a senior capstone course or project is required, so if you completed a report or a research paper for your capstone, chances are it had many of the same elements as a business plan.

Business plans account for all the major areas that you need to consider when going forward in any opportunity. Whether you are planning a vacation, planning a wedding, going onto graduate school, writing your research proposal, or starting a business, if you can account for the following five areas and do your best to anticipate concerns, problems, or shortcomings in every area, you will be well served.

Remember, business plans are dynamic documents, and should be revisited regularly. Information changes, funding changes, personnel changes... these things may alter one or several elements of the plan.

With that said, let's talk a little bit about each section of a business plan and then look at how you can use this to put together a plan for any upcoming endeavor.

The Framework of a Business Plan

Depending on the context or purpose of your business plan, each of the following sections that comprise the framework of a business plan may have a slightly different name.

As you are working your way through your business plan, keep in mind that what you are doing is accounting for as many of the controllable variables as possible. Of course, there will always be the unforeseen and the uncontrollable; this plan is just to help you with any and all of the variables that you should be able to identify in advance. Having a plan in place for the variables you can control will make life considerably easier when the unexpected occurs.

Business Idea

In your business plan, this is the "big idea". That is, what is it exactly that you want to do? Are you bringing a product or service to the market? If so, is it brand-new, or are you refining it? Your business idea should be stated clearly in one or two sentences. You should be able to go on to describe it in detail, and you should also be able to articulate your idea very succinctly so that people can understand it easily, no matter how technically or technologically complicated it is. You need to be able to sell your idea to the general public.

As an expert on your idea (which you should be!), you will want to describe every aspect of it in stunning detail—but save that for the next section in your business plan. For now, write your business idea clearly and succinctly, and then run it by your friends, a parent or two, or another trusted advisor or mentor. If they can repeat it back to you in a way that makes sense, they have understood your idea, and you have clearly communicated what you want to do. Good job! If they can't, simplify further and try again.

Market Analysis

Your market analysis is where you describe the lay of the land. Who are your competitors, what are their products and capabilities, and how do you anticipate outselling or outperforming them? You can have subsections here that describe the differentiation of your product, whether incremental or a step function. This is where you also include potential changes in market trends and the history of the marketplace.

You want to give a thorough review of the area in which you are entering. This demonstrates to potential investors that you know your business, any potential threats, and where your opportunities lie.

Do not sugarcoat the challenges or difficulties that will stand in the way of the realization of the business plan. Doing so will only demonstrate to your investors that you have a gilded understanding of reality. We are all inclined to confirmation bias because we all love and are in love with our ideas, so be wary of this. Simultaneously, don't overstate the difficulties or your shortcomings, either, as you don't want to unnecessarily scare off your investors.

The better a job you do of assessing the marketplace and the lay of the land dispassionately, the better you will be able to anticipate difficulties.

Remember that difficulties do not equal impossibilities, so do not be afraid of them; just give them a good, hard, honest look so you know exactly what you're dealing with. Your investors will appreciate this levelheaded approach and be more inclined to support your business endeavors.

People

Who are the people you need to be involved for your business to succeed? There will be some obvious groups, like your customers, but you will need to define who exactly you are targeting in the market space for your product or service. You will also have other stakeholders, depending on the type of business you are planning. You could need board members, certainly employees, a business mentor or two, allies in the local government (if you will be needing a building permit or a variance), specialists from different agencies (if you are looking for an environmental impact survey), an accountant or a lawyer, and a marketing or PR firm to help with your business launch. The list goes on, and there are many professionals who will play large, medium, and small roles in your business' development, launch, and growth, so take the time to think about all the different players you will need, and then start penciling out the relationships you want to build and develop.

People are vital to any business, both on the customer side and the business side, so it should be your priority to develop a solid understanding of who these people are and how to engage with them so your business will succeed.

Mentors will help you build this list, and will oftentimes make introductions for you, either in person or virtually. These formal and informal mentors are invaluable. If you do not have one, you can start with your local chamber of commerce for ideas on who to meet and how to meet them.

Money

You have to spend money to make money, and in any new business, it can seem that the spending is happening a lot more than the making. For this reason, if you are starting a new business, you may have to take out a business loan, depending on your circumstances. If this is the case, you will want to demonstrate an understanding of all three major financial statements (the balance sheet, income statement, and cashflow statement). If you have not actually sold anything yet, your statements should reflect reasonable projections (though what is "reasonable" is something that is widely debated, since new business owners and entrepreneurs are often optimistic about their sales, while bank loan officers are, inversely, realistic, if not pessimistic).

In addition to money for the resources needed to make your product, you will need money for rent and utilities, personnel, taxes, basic business office needs, advertising, a website, and professional services (as mentioned in the "People" section previously).

There are lean methods for starting new businesses, and these will usually work in your favor, so when you are putting together your business plan, pencil out every conceivable detail that you might need to spend money on, and then go back and determine if it's worth using your business loan to cover these costs, or if you can cut them until you have income. You do not want to take on any more debt than is absolutely necessary.

Timeframe

The timeframe in your business plan can be anything from when you expect your business to be profitable, to your first product or service launch, to just about any other milestone. Usually, however, timeframes are tied to when your business will be "in the black".

It is not expected that you will be turning a profit right out of the gate—many businesses take three to five years to be profitable, so banks will expect this—but as your first step, you will need to be able to show that you are making enough money to pay your bills. If you are only able to pay your bills because you are constantly infusing cash (which is just additional debt), then your business is not profitable.

That isn't to say that you won't, as a business owner, possibly end up putting your own money into your business. This is a reasonable expectation. However, it is important that you decide ahead of time how much you are willing to invest and at what point you will stop, not when you are in the throes of pumping money into a business that may or may not be viable.

Having a timeframe, with decision points penciled out, before you get started, will also help to decrease your emotional responses and maintain your level head when you're making potentially fraught decisions, like whether you should maintain your investment, invest in other products or businesses, or even pull the plug entirely.

There are scores of books on this subject, and I strongly encourage you to read several. I also recommend that you talk to actual business owners. You will learn a great deal more from visiting established local businesses than just reading. Local business owners will be able to tell you about the nuances of your potential market, and, depending on their level of openness, they may also tell you about how long it took them to run in the black and about the milestones that mattered to their business in the end. At a minimum, they will certainly understand the local business environment, which may prove insightful.

Applications

Why do I spend all this time talking about business plans in a book dedicated to study habits for STEM undergraduates?

Most students that I meet in STEM disciplines are, in some form or another, the planning type, and with a little foresight, a pencil, and your engineering paper (or any notebook, for that matter), you can use the aforementioned business plan for just about any endeavor. If you account for all the major topics in a business plan (whether you are building an actual business, drafting your proposal for doctoral work, planning your political campaign, or even planning a wedding), then you will have accounted for ninety-five percent of what's to come.

The more you plan so that you can account for the controllable variables, the more creative energy you will be able to spend on addressing uncontrollable variables when they arise—and that is what sets the big dogs apart from the puppies.

To illustrate, here are a couple of examples of how to use a business plan for any situation, so that when you go forward, you can have an idea of what you will be up against (which will increase your chances of success dramatically).

Drafting your proposal should look something like this: the background or marketplace description is your literature review (demonstrating that you have sufficiently surveyed all materials that are relevant to your study question); the "big idea" is your testable hypothesis; the people involved are your committee members, laboratory directors, or anyone involved in helping you acquire the needed resources; the money will be how you intend to pay for your research;[32] and the timeframe is when you intend to be finished (I recommend no longer than four years from start to finish).

If you are running a political campaign, then your sample business plan would look like this: your "big idea" will either be your slogan or your two or three platform points that you are running on; the people involved will both be your inner circle who help advise you on a variety of subjects (like policy and marketing) and your voters; for money, you will need a budget to pay for the marketing materials and possibly staff (if volunteers are not enough); and the timeframe will be starting on Election Day and working backward to determine everything you want to get done (without leaving all your efforts until two weeks before the Election Day).

For your wedding: the "big idea" should be that you are getting married

---

[32] If you are not funded, that is. If you are, then you will need to know how you are going to allocate such funds.

(and presumably you have your intended spouse already picked out!); the people involved will be everyone, from the guests at your wedding, to the officiant, to who will be making your cake, to who will be delivering flowers, to who will be assisting with invitations, and, of course, your wedding party; the money will be, again, your budget and how you intend to allocate the funds based on your wedding priorities (they grow expensive quickly, so set your budget and hold firm!); and the timeframe will be when you intend to finish certain aspects of your wedding planning and decision making, up to and including the big day.

As you can see, business plans truly are applicable for just about any idea that you want to take from a twinkle in your eye to a fully executed reality. Pencil out your plan, account for the controllable variables, and go forward knowing you have done everything in your power to allow you to you have the mental bandwidth to contend with the unknowns.

Recommended Reading

- *The Martha Rules: 10 Essentials for Achieving Success as You Start, Build, or Manage a Business* by Martha Stewart.
- *The CEO Next Door: The 4 Behaviors that Transform Ordinary People into World-Class Leaders* by Elena L. Botelho, Kim R. Powell, and Tahl Raz.
- *Managing Oneself* by Peter F. Drucker.
- *Total Recall: My Unbelievably True Life Story* by Arnold Schwarzenegger.

Strategies for Success: Business Plans

- Having a business plan for your "big idea" will significantly increase your chances of success.
- Define your "big idea".
- Know who the people that you need are.
- Know how much money you have, where it will be coming from, and

what your deadline is.

- Do a thorough job of your background research (market analysis).
- Know that you can apply the business plan basics to any idea you have in life, and it will get you ninety-five percent of the way to the finish line.

End Note

Quite often, a genius is simply someone who has done all their homework.
—THOMAS EDISON

Congratulations! You've made it, my friend. You've made it to the end of this book, to the end of your class, or the end of your degree program. You worked hard, you practiced problems and studied the material, you took care of your health, and you made friends.

What's next?

First, take a minute to catch your breath. Quite often, when we finish something big ("big" is however we define it, by the way), there is a bit of a lull that follows right on the heels of our accomplishment. You spent tremendous time and energy focusing on the goal and seeing it through to the finish line, and now it is done, your focus and attention are not singularly trained on the end point. It is normal to feel a bit untethered at this stage, so give yourself a reward, such as a hike in a national park, a night out with your friends, or a day trip out to that place you've always wanted to visit but could never afford the time to before now. Celebrations do not need to cost a lot of money, so get creative! You've earned it.

Next, look at the plan you sketched out to accomplish your goal. Do you have additional courses you need to take, or are you still planning your after-college move? Every time we move through a plan to finish a goal, we have the opportunity to do it better. We review, we evaluate, we audit, we adjust, we improve, we build templates. We optimize. Did you take too many labs in one semester and the time commitment nearly crushed you? Good to know. Plan a more even distribution of lab requirements for your upcoming semesters. Did you wait too far into the semester to start your semester-long project and end up with a mediocre grade? Good: draw up a plan and stick to it for the next go around. Did you spend too much money this semester and need to dial back in the next? Good: look at your expenses and determine if these were necessary and unplanned, or if they were unnecessary and just plain expensive. This practice of conducting a review—an honest assessment of what worked and what did not—and defining a plan for doing it better next time will not only increase your proficiency, but will also minimize mistakes and wasted time and effort, and produce a better outcome.

Lastly, a personal request: invite others to do what you have done. Your hard work is helping or has helped you to finish something very few other people have ever considered doing (a degree in STEM), and the facts are, we need more students to study the technical fields. No, not everyone will be interested in this, but everyone should know these courses are doable, these degrees are attainable, and these careers are remarkable.

After seeing my grades and observing all that I had done in high school, my high school guidance counselor recommended one of two careers for me: teaching or nursing.[33] I had no interest in either field, and I knew I was going to study international business, so I dismissed her comments and forged on. In retrospect, both of those are fine careers—my Grandma Genevieve was a nurse, and her mom (my Great-Grandma Rosina) was a teacher who changed the way special education was done in South Dakota—but these were still not the careers for me.

As I neared the end of my first bachelor's degree, I mentioned to one of my business professors that I had really wanted to be an astronaut, and wished I had studied math in college. She looked at me and asked, "Why not now?" I eyed her with dubiousness. I could not comprehend starting a math degree at that "late" stage; I was nearly twenty-two! It was time for me to graduate and start working.

...Right?

The humor is not lost on me that my fixed mindset when I thought I was "too old" at twenty-two transformed into a growth mindset at thirty-two— the age I finally relented and went back to college to study math, science,

---

[33] Believe it or not, this recommendation was given to me in the mid-1990s, not the mid-1950s.

and engineering.

If you've read Carol Dweck's book *Mindset*, you'll know achievement is less about your inherent smarts and more about your willingness to dig in and work hard. If you've read Angela Duckworth's book *Grit*, you'll know that resilience is key. You're going to take your lumps, so accept that and move forward.

Who ultimately invited me to study STEM? Certainly, there was my friend who suggested the petroleum engineering degree, the woman in admissions who told me which courses I needed to take to transfer in as a post-baccalaureate, and my first advisor at MT Tech who signed me up for eighteen credits and had every confidence I could do the work. But the first would have to be my sister, Kendra.

She is four years younger than me, and she knew from an early age that she was going to work in the space program. At one point, she read a book about the space shuttle *Challenger*, which exploded because the engineers' warnings had not been heeded. We lost irreplaceable astronauts because leadership rushed to make a deadline.

She was not going to let that happen again.

Whenever I came home from college, I saw Kendra studying hard. She had her books and notepaper scattered on her bed while she cranked through problem sets. I went with her to look at universities, and when she decided, I would go visit her in Boston.

She continued that hard work with her ROTC commitments and her homework, and she managed to work in a French minor with a semester spent studying abroad.

I watched her work *hard* to do all of it.

This is why I am asking you to invite the next generation to study science, technology, math, and engineering. Sometimes, we need an outright invitation, like my petroleum engineering friend gave to me: "I think you would be good at this. Have you thought about studying STEM?" Sometimes, we just need to see someone who will let us know there is a whole world of opportunity out there, like my sister did.

When you have the chance to invite others to join you in STEM courses, degrees, and careers, be honest. Let them know it is hard, but that it can be done. Share your story of failure and triumph. Give them a sense of the thrill you have when you solve that intractable problem. Go back to your old high school or middle school and talk to the students. Take examples of your work and show what you can do. Offer to judge at a science fair, or give a guest lecture for one of their math or science classes.

There are still so many bits and pieces that I would love to share with you, but you need to get on with that homework. The day will come when you write your own story, so when you do, make sure you include lessons you've learned. You never know who needs to hear exactly what you have to say.

Your STEM Study Partner,

Dr. Scyller Borglum

Acknowledgments

Learn from the mistakes of others. You can't live long enough to make them all yourself.
—ELEANOR ROOSEVELT

My husband, Tim, whether he wanted to or not, discussed every topic in this book with me endlessly. Tim, this book would not be what it is without your support, experience, and perspective.

Trina Mata Barr: you asked the question, "Why not study petroleum engineering?"

Arletta Tompkins: who took my call and provided me with the basics for how to prepare and what to do to apply for their engineering program.

Lana Petersen: who answered the phone when I called MT Tech to inquire about applying.

Mary North Abbott: who took me in as a transfer student with virtually no background in engineering, and ensured I took every class I needed to succeed. She kept me on as her teaching assistant, which allowed me to learn all that cannot be taught in a book (even this one!). She also invited me to teach my first workshop to a group of students in desperate need of organizational help, and that little seminar was what started this entire *Study Habits* adventure!

Dave Reichhardt: thank you for teaching me how to write a scientific paper and for reviewing my work *ad nauseum*.

Burt Todd, my master's thesis advisor: I cannot thank you enough for the wisdom you imparted and the patience you showed me in every capacity (so many in which I decidedly did not deserve).

Dan Soeder: for inviting me to apply for the ORISE Fellowship and work at NETL for the summer (and for inviting me to coauthor my first book with him).

Acknowledgments

Lisa Carlson: for inviting me to work with the Student Success Center on SD Mines' campus.

Matt Hanley: for his expertise in postgraduation opportunities.

Isaac Egermeier: for his considerable tutelage in coding.

Faye, my editor: thank you so much for your willingness to work through drafts of writing. My readers appreciate you more than they will ever know.

Hayley, my publisher: I am so proud to be an author in your publishing company. You built something remarkable.

Kendra, my sister and inspiration: thank you for your example!

Last but not least: my mom and dad. You gave us the space and freedom to grow. Thank you. I hope you see your fingerprints all over the stories and lessons in this book.

Appendix A
SJB Breakfast

After years of eating my Scyller Breakfast (or the "SJB Breakfast") in front of fascinated/horrified onlookers (and after receiving a couple of requests), I decided this recipe warranted its own section.

Initial considerations when making the SJB Breakfast:

- I practice intermittent fasting, so by the time I eat my first meal of the day (around 10AM or 11AM), I am famished.

- It is too much to make every single day. Instead, I line up four to five glass containers and, assembly line fashion, prep the entire week's breakfast at once, usually on a Sunday afternoon. We need to eat every day, but who has the time to actually prepare food every day?

- After eating my SJB Breakfast, I drink a liter of water. Do this, and you will not be hungry for hours. You will be able to motor through your day cheerfully, nourished and productive.

- The initial purchase of the ingredients can be expensive, but you are only using a little bit of each at a time, and most of them are non-perishable. Store according to package instructions, and your breakfast accoutrement will be around for a long time.

- I assemble all at once, but wait to stir until the day I eat my SJB Breakfast.

- At least one item from every line is used for each batch. Depending on availability, I will use what is in season, on sale, or in my cupboards. While the following instructions endeavor to be comprehensive, do not put every single item in your containers, because who can afford that?

- As for the measurements, those are an estimate. I find it vexing to read recipes that do not provide even an idea of scale, so that is the purpose of the recommended measurements. Do not actually bother with cups, tablespoons, or teaspoons. Just eyeball it.

Prep

1. Make the steel cut oats, if that will be your base ingredient. (Rolled oats or muesli do not need to be premade. The milk soaking into them overnight will soften the cereal.)
2. Line up 4-5 glass containers on your countertop.
3. Assemble the SJB Breakfast accoutrement.

Ingredients

- ¾ C muesli (can be put in container raw) or steel cut oats (needs to be prepared ahead of time, according to instructions. I use Bob's Red Mill for both).
- 1 C frozen fruit (blueberries, strawberries, raspberries, blackberries, etc.).
- Small handful dried fruit (raisins, dates, cranberries, cherries (my favorite!), goji berries, etc.).
- Handful of nuts (walnuts, pecans, cashews, Brazil nuts, etc.).
- 2 Tbsp protein (hemp protein, ground flax, etc.).
- 1 Tbsp chia seeds, sliced almonds, unsweetened coconut flakes, etc.
- 1 Tbsp maple syrup, agave nectar, honey, etc.
- Salt (regular table salt, Himalayan Sea salt, kosher salt... it does not matter).
- Cinnamon (be liberal and enjoy it!).
- A little pat of butter.[34]

Instructions

1. Put all the ingredients in the glass containers.

---

[34] My seventh- and eighth-grade home ec teacher, Mrs. Jacobsen, told us we all need a little fat in our diets for smooth, younger looking skin. I believed her then, and I believe her now. Use real butter.

2. Cover the entire assemblage with kefir or almond milk.
3. Put the lids on your glass containers.
4. Stack neatly in your refrigerator.
5. Know that your week is off to a fantastic start.

Be warned: when you whip out your SJB Breakfast, people will wrinkle their noses at you and ask, "What is that?!" You can ignore or respond to the philistines (your choice), but expect the question either way. If you have a blueberry and hemp protein combination, it will be a ghastly gray color (think concrete or an extraterrestrial conglomerate), but do not worry, your breakfast is still delicious!

The SJB Breakfast can be heated in the microwave. This will help with melting your little pat of butter, and will bring a warm smile to your face on a cold Dakota day!

Appendix B
Books Worth Reading

The list of books provided here will be a solid start to your personal library. I have a strong commitment to only recommending books that I have already read, so I can assure you that each book is a good read and will support you in your academic and future professional journey. Use these books and authors as personal mentors who can be accessed day or night.

This list is a sampling of good books, and is by no means comprehensive. You will have your own titles to add.

- *The 5 Second Rule: Transform Your Life, Work, and Confidence with Everyday Courage* by Mel Robbins.
- *24/6: A Prescription for a Healthier, Happier Life* by Matthew Peterson, with Eugene Sleeth.
- *The 80/20 Principle: The Secret of Achieving More with Less* by Richard Koch.
- *American Icon: Alan Mulally and the Fight to Save Ford Motor Company* by Bryce G. Hoffman.
- *Astrophysics for People in a Hurry: Essays on the Universe and Our Place Within It* by Neil Degrasse Tyson.
- *Atomic Habits: An Easy and Proven and Way to Build Good Habits and Break Bad Ones* by James Clear.
- *The Autobiography of Benjamin Franklin* by Benjamin Franklin.
- *Becoming Odyssa: Adventures on the Appalachian Trail* by Jennifer Pharr Davis.
- *Basic Black: The Essential Guide for Getting Ahead at Work (and in Life)* by Cathie Black.
- *The Big Sister's Guide to the World of Work: The Inside Rules Every Working Girl Must Know* by Marcelle DiFalco and Jocelyn Greenky Herz.

⊐ *The Boy Who Harnessed the Wind: Creating Currents of Electricity and Hope* by William Kamkwamba and Bryan Mealer.

⊐ *Broad Band: The Untold Story of the Women Who Made the Internet* by Claire L. Evans.

⊐ *Business Etiquette for Students and New Professionals* by Mary Crane.

⊐ *Call Sign Chaos: Learning to Lead* by Jim Mattis and Bing West.

⊐ *The CEO Next Door: The 4 Behaviors that Transform Ordinary People into World-Class Leaders* by Elena L. Botelho, Kim R. Powell, and Tahl Raz.

⊐ *Cheaper by the Dozen* by Frank B. Gilbreth Jr. and Ernestine Gilbreth Carey.

⊐ *The Coddling of the American Mind: How Good Intentions and Bad Ideas Are Setting Up a Generation for Failure* by Jonathan Haidt and Greg Lukianoff.

⊐ *The Coming Plague: Newly Emerging Diseases in a World Out of Balance* by Laurie Garrett.

⊐ *Daily Rituals: How Artists Work* by Mason Curry.

⊐ *Daily Rituals: Women at Work* by Mason Curry.

⊐ *The Disappearing Spoon: And Other True Tales of Madness, Love, and the History of the World from the Periodic Table of the Elements* by Sam Kean.

⊐ *Discipline Equals Freedom: Field Manual* by Jocko Willink.

⊐ *Discipline is Destiny: The Power of Self-Control* by Ryan Holiday.

⊐ *The Elements of Style* by William Strunk Jr. and E.B. White.

⊐ *The Emperor of All Maladies: A Biography of Cancer* by Siddhartha Mukherjee.

⊐ *Endurance: Shackleton's Incredible Voyage to the Antarctic* by Alfred Lansing.

⊐ *Everything is Figureoutable* by Marie Forleo.

⊐ *Executive Presence: The Missing Link Between Merit and Success* by Sylvia Ann Hewlett.

⊐ *Fortitude: American Resilience in the Era of Outrage* by Dan Crenshaw.

⊐ *Getting Things Done: The Art of Stress-Free Productivity* by David Allen.

⊐ *Grit: The Power of Passion and Perseverance* by Angela Duckworth.

⊐ *Guns, Germs and Steel: The Fates of Human Societies* by Jared Diamond.

⊐ *Hidden Figures: The Story of the African-American Women Who Helped Win the Space Race* by Margot Lee Shetterly.

⊐ *How to Work a Room: The Ultimate Guide to Making Lasting Connections—In Person and Online* by Susan RoAne.

⊐ *How to Stop Worrying and Start Living: Time Tested Methods for Conquering Worry* by Dale Carnegie.

⊐ *How to Win Friends and Influence People* by Dale Carnegie.

⊐ *How Will You Measure Your Life?* By Clayton Christensen.

⊐ *Humble Pi: A Comedy of Maths Errors* by Matt Parker.

⊐ *The Ideal Team Player: How to Recognize and Cultivate The Three Essential Virtues* by Patrick M. Lencioni.

⊐ *The Immortal Life of Henrietta Lacks* by Rebecca Skloot.

⊐ *The Invention of Nature: Alexander von Humboldt's New World* by Andera Wulf.

⊐ *Kon-Tiki: Across the Pacific by Raft* by Thor Heyerdahl.

⊐ *The Last Place on Earth: Scott and Amundsen's Race to the South Pole* by Roland Huntford.

⊐ *Let Your Mind Run: A Memoir of Thinking My Way to Victory* by Deena Kastor.

⊐ *The Little Engine That Could* by Watty Piper.

⊐ *Longitude: The True Story of a Lone Genius Who Solved the Greatest Scientific Problem of His Time* by Dava Sobel.

⊐ *Lord of the Flies* by William Golding.

⊐ *The Magic of Thinking Big* by David J. Schwartz.

⊐ *Managing Oneself* by Peter F. Drucker.

⊐ *The Martha Rules: 10 Essentials for Achieving Success as You Start, Build, or Manage a Business* by Martha Stewart.

⊐ *Mindset: The New Psychology of Success* by Carol S. Dweck.

⊐ *My American Journey: An Autobiography* by Colin Powell.

⊐ *Never Eat Alone: And Other Secrets to Success, One Relationship at a Time* by Keith Ferrazzi and Tahl Raz.

⊐ *On Writing: A Memoir of the Craft* by Stephen King.

⌐ *The One Thing: The Surprisingly Simple Truth Behind Extraordinary Results* by Gary Keller, with Jay Papasan.

⌐ *The Power of Habit: Why We Do What We Do in Life and Business* by Charles Duhigg.

⌐ *The Pursuit of Happyness* by Chris Gardner.

⌐ *Quiet: The Power of Introverts in a World That Can't Stop Talking* by Susan Cain.

⌐ *Rich Dad, Poor Dad* by Robert T. Kiyosaki.

⌐ *The Richest Man in Babylon* by George Clason.

⌐ *Sam Walton: Made in America* by Sam Walton, with John Huey.

⌐ *Sea Change: A Message of the Oceans* by Sylvia Earle.

⌐ *The Soul of an Octopus: A Surprising Exploration Into the Wonder of Consciousness* by Sy Montgomery.

⌐ *The Telomere Effect: Living Younger, Healthier, Longer* by Elizabeth Blackburn, PhD and Elissa Epel, PhD.

⌐ *Total Recall: My Unbelievably True Life Story* by Arnold Schwarzenegger.

⌐ *The War of Art* by Steven Pressfield.

⌐ *West With The Night* by Beryl Markham.

⌐ *What Makes Olga Run? The Mystery of the 90-Something Track Star and What She Can Teach Us About Living Longer, Happier Lives* by Bruce Grierson.

⌐ *When Books Went to War: The Stories That Helped Us Win WWII* by Molly Guptill Manning.

⌐ *Why We Sleep: Unlocking the Power of Sleep and Dreams* by Matthew Walker.

⌐ *Write it Down, Make It Happen: Knowing What You Want and Getting It* by Henriette Klauser.

⌐ *Zero: The Biography of a Dangerous Idea* by Charles Seife.

Appendix C
FAQs

Q1: My birthday is coming up and I know exactly what I want. Should I ask for a birthday pony?

A1: No. You can buy your own pony after you graduate. You should be asking for what you really need: a gift card to the bookstore, new underpants or dress socks, an interview suit, a subscription to your hometown newspaper (or a paper of your choice), or running shoes.

Q2: I'm majoring in mechanical engineering, but I'm also really interested in Egyptology. Should I change majors?

A2: Continue majoring in mechanical engineering and take a minor (or possibly a double major, if you can realistically manage it) in Egyptology. College should be fun, and layering on a minor for a side passion adds to the experience. Whether you like music or French literature or medieval history, go for it. You never know where it may lead. Although I did not plan it that way, my Scandinavian studies minor contributed significantly to my being awarded a Fulbright Scholarship. Your engineering (or other STEM) degree will land you the career you enjoy so you can continue your armchair interests and hobbies.

Q3: My pajamas are really comfortable, and I would like to stay in them all day to save laundry. Is this a good idea?

A3: No. You need to be out of your pajamas by 8AM. Put on day clothes (jeans and a T-shirt, or a polo or sweater, for example), not basketball shorts or yoga pants. These are pajamas pretending to be socially acceptable day clothes. Now, get after your day.

Q4: I've been smoking up all semester, and now the company who hired me wants a hair (follicle) test for drugs. I shaved everything. This plan will work, right?

A4: No. You need to be able to pass a drug test when they ask for it. Companies offering competitive positions are not going to wait for your hair to grow out so you can pass your drug test. Plus, everyone on campus preparing to start their summer internships or postgraduation full-time work also just submitted for a hair test, and it won't take a rocket scientist to figure out why you showed up for class shorn from head to toe.

Q5: I stayed up all night studying for my final exam and then slept through my alarm. What do I do?

A5: Own it. Go immediately to your professor, apologize profusely, and take responsibility. Do not make up some fantastical story with exaggerated claims. You may be surprised at their generosity. Some might have you take it then and there, and some might schedule a make-up exam. If they do not allow you to sit for the final, thank them for their time (this was your mistake, after all, not theirs) and accept it. As disheartening as it feels, it isn't the end of the world. Take this as the lesson that it is and do not repeat it.

Q6: I'm completely overwhelmed and stressed out. I have so much to do, but I really want to take a nap. What should I do?

A6: Take a nap. Seriously. Crawl under your desk in the study room, go out to your car in the parking lot, or crash on your sofa. Set your timer for twenty-five minutes, grab your earplugs for silence, put a T-shirt over your eyes to block out the light, and sleep until your alarm goes off. Then, get up (and make a cup of coffee, if that is your thing) and write out a to-do list. When you are this stressed out, don't bother trying to arrange it in the most effective way: just start listing out everything you need to do. If your brain is in gridlock, it does not matter which item you start with (the first, shortest, easiest, or most accessible); just pick one item and get it done. Then, go after another one. A few items in, your mental gridlock will start freeing up, and you'll be able to proceed in a more organized manner.

Q7: I recently met someone who makes decisions based on neither logic nor reason. May I conduct mini social experiments on them and track their responses?

A7: No. As tempting as it may be to run experiments on our fellow humans, it is not allowed in any form without signed informed consent. In fact, conducting said experiments could be considered gaslighting your fellow human, and suffice to say, this is frowned upon. Don't be antisocial. Go make friends with this unusual person and appreciate your differences as people.

Q8: I have a rock-solid morning routine (that I start the night before!), but an early meeting/appointment/flight. How do I manage my morning?

A8: You have two options: abbreviate each of your morning commitments and know that something is better than nothing (completely valid approach), or, if you are certain you can pick it up later in the day, postpone some of the activities you'd normally reserve for the morning to later in the day. I frequently have early flights and move my training to the evening when I arrive at my hotel. Your routine may be jeopardized in some way if you have a delayed or missed flight, however. If this happens, shrug it off and get right back after it the next morning.

Q9: I am closing in on the end of my graduate degree and need every minute to concentrate. What do I do about my club involvement and extracurricular activities?

A9: Deena Kastor has a marvelous description in her book, *Let Your Mind Run*, about coming down the homestretch of a long race. She whips off her sunglasses and tosses them to the side as she focuses everything in her body, mind, and spirit on her final goal. For her strong finish, she does not even want to carry the extra weight of sunglasses!

On occasion in our lives, we need to strip everything out that is not absolutely necessary to get to our goal. This "hyperfocus" technique is not a reason to jettison all the good habits you built up, but a "boiling down" of only the most essential energy expenditures for a specific time period to get you to the finish. When I am that focused, I maintain my Bible study and prayer time, physical training, personal hygiene, and good food and sleep. Nothing else. Once the goal is accomplished, pick back up again with your extracurriculars, and enjoy.

About the Author

Dr. Scyller Borglum currently serves as Vice President of Energy in Underground Storage at WSP USA. Scyller served as South Dakota State Representative from 2018–2020 before returning to the technical field as a subject matter expert for underground storage.

Dr. Borglum formerly conducted doctoral research in geology and geological engineering in the RESPEC Geomechanical Laboratory and as an ORISE Fellow at the National Energy Technology Laboratory in Morgantown, WV. During this time, Scyller also coauthored the textbook *The Fossil Fuel Revolution: Shale Gas and Tight Oil*.

Scyller started in the energy industry as a petroleum engineer, working up and down the Rocky Mountain Front and in North Dakota. Now, Scyller advises clients, local, state, and federal agencies, think tanks, and elected officials (both in the U.S. and abroad) about storing energy subsurface for energy security and accessibility.

Dr. Borglum is a passionate advocate for bringing new students into STEM studies, particularly energy production, generation, development, storage, and deliverability.

http://linkedin.com/in/sjborglum

www.ingramcontent.com/pod-product-compliance
Lightning Source LLC
Chambersburg PA
CBHW021923190326
41519CB00009B/891